Applying Regression & Correlation

Applying Regression & Correlation

A Guide for Students and Researchers

JEREMY MILES AND MARK SHEVLIN

SAGE Publications
London • Thousand Oaks • New Delhi

ISBN 978-0-7619-6229-8 (hbk)
ISBN 978-0-7619-6230-4 (pbk)
© Jeremy Miles and Mark Shevlin 2001
First published 2001
Reprinted 2002, 2003 (twice), 2004, 2005, 2006, 2013

Apart from any fair dealing for the purposes of research or private study, or criticism or review, as permitted under the Copyright, Designs and Patents Act, 1988, this publication may be reproduced, stored or transmitted in any form, or by any means, only with the prior permission in writing of the publishers, or in the case of reprographic reproduction, in accordance with the terms of licences issued by the Copyright Licensing Agency. Inquiries concerning reproduction outside those terms should be sent to the publishers.

SAGE Publications Ltd
1 Oliver's Yard
55 City Road
London EC1Y 1SP

SAGE Publications Inc
2455 Teller Road
Thousand Oaks
California 91320

SAGE Publications India Pvt. Ltd
B1/I Mohan Cooperative Industrial Area
Mathura Road
New Delhi 110 044

SAGE Publications Asia-Pacific Pte Ltd
3 Church Street
#10-04 Samsung Hub
Singapore 049483

British Library Cataloguing in Publication data

A catalogue record for this book is available from the British Library

Printed on paper from sustainable sources

Typeset by Mayhew Typesetting, Rhayader, Powys
Printed and bound in Great Britain by
Ashford Colour Press Ltd

Contents

Preface vii

PART I I NEED TO DO REGRESSION ANALYSIS TOMORROW

1 **BUILDING MODELS WITH REGRESSION AND CORRELATION** 1

What are models? • Least squares models • A very simple model • The standard error of the mean • Modelling relationships • The standard error and significance of parameter estimates • Standardised estimates • Looking more at correlations • Correlations and scattergraphs • Correlations and variance • Correlations and size • Notes • Further reading

2 **MORE THAN ONE INDEPENDENT VARIABLE — MULTIPLE REGRESSION** 27

Introduction: multiple regression in theory • What's multiple regression all about? • Multiple regression in practice • R and R square • Adjusted R square • Analysis of variance (ANOVA) table • Coefficients • Variable entry • Hierarchical variable entry • Methods of variable entry • Note • Further reading

3 **CATEGORICAL INDEPENDENT VARIABLES** 40

Introduction • Categorical data: a special case • The t-test as regression • ANOVA as regression • Coding schemes for categorical data • Notes • Further reading

PART II I NEED TO DO REGRESSION ANALYSIS NEXT WEEK

4 **ASSUMPTIONS IN REGRESSION ANALYSIS** 58

Introduction • Assumptions about measures • Levels of measurement • Conservative interpretation of assumptions • A more liberal approach • Assumptions about data • A bit about normal distributions • Univariate distribution checks • Outliers and the mean • Normal distribution • Detecting and dealing with non-normality • Calculation-based methods • Skew and kurtosis • Outliers • Dealing with outliers, skew and kurtosis • Dealing with outliers • Effects of univariate skew and kurtosis • Multivariate distributions • Assumption 1 • Assumption 2 • Assumption 3 • Assumption 4 • Time-series designs • Clustered sampling designs • Notes • Further reading

5 ISSUES IN REGRESSION ANALYSIS 113

Causality • Association • Direction of causality • Isolation • The role of theory in determining causation • Sample size • Why should we worry about sample sizes? • Rules of thumb • Power analysis • Collinearity • What is collinearity? • Detecting collinearity • Dealing with collinearity • Measurement error • Notes • Further reading

PART III I NEED TO KNOW MORE OF THE THINGS THAT REGRESSION CAN DO

6 NON-LINEAR AND LOGISTIC REGRESSION 136

Non-linear regression • Linear and curvilinear relationships • Generating a curve • Carrying out non-linear regression • An example of non-linear regression • Logistic regression • The case of the dichotomous dependent variable • The logit transformation • Using the logit: logistic regression • An annotated example of logistic regression • Hierarchical logistic regression • Polynomial logistic regression • Further reading

7 MODERATOR AND MEDIATOR ANALYSIS 165

Introduction • Moderator analysis • Two categorical variables • Categorical and continuous variables • Two continuous predictors • Mediator analysis • Example of mediation • Some concluding points on moderation and mediation • Note • Further reading

8 INTRODUCING SOME ADVANCED TECHNIQUES: MULTILEVEL MODELLING AND STRUCTURAL EQUATION MODELLING 192

Multilevel modelling (MLM) • Algebraic formulation • Hierarchies everywhere • Even more hierarchies • Structural equation modelling • Why use SEM? • Identification • Latent variables • Estimation in SEM • Model testing • Structural models • Programs for MLM and SEM • MLM software • SEM software • Notes • Further reading

Appendix 1	**Equations**	216
Appendix 2	**Doing regression with SPSS**	228
Appendix 3	**Statistical tables**	237
References		245
Name index		249
Subject index		251

Preface

We are not statisticians.

There. We just wanted to clear that up at the start. We have met a lot of statisticians in our time, and we have a great deal of admiration for them. They are more knowledgeable than we are, they seem to understand things that we don't understand, and they are usually cleverer than we are.

We spend a lot of our time doing what might be termed 'statistics'. We do statistical analyses, we advise people on statistics, and we investigate the limitations of statistical techniques. But we are not statisticians. We are psychologists.

So the reader may ask, 'Why have you written a book on statistics?' This is a good question to ask, but it isn't the right question to ask. If you glance at this book, or flick through it, perhaps having seen it on a friend's bookshelf, or in a library or bookshop, you could well put it down and believe that it is a book on statistics. It has got numbers, equations (though not many) and Greek symbols in it, so we don't mind if, at first glance, you think it is a book on statistics.

But we hope that someone who sits and reads a larger part of the book will realise that it is not a book on statistics, rather it is a book on psychology. Psychology is the scientific study of behaviour – usually human behaviour. It is about trying to determine the causes of behaviour, and uncovering the myriad of complex processes that are going on inside people's heads that cause them to do the things they do. But if we are to represent and decode these processes, we need to have a system that we can use, and the system that is used by the vast majority of psychologists is numbers. We try to measure behaviour, and turn it into numbers. Then we can search for patterns within these numbers, but we should always remember that we are not interested solely in the numbers. In fact, we are probably not interested in the numbers at all. We are interested in what those numbers represent, and we should never forget that those numbers represent behaviour. And if we are going to uncover the intricacies of the psychological process, the only weapon that we have in our arsenal is behaviour (barring biological interventions). And if we want to uncover the intricacies of behaviour we need to use appropriate and adequate statistical techniques to uncover the patterns. Regression analysis represents a useful, flexible tool to be applied to this analysis in an attempt to uncover patterns. And that is why this book is about psychology even though it may have the appearance of being about statistics.

We would like to acknowledge the assistance of a great many people who have read and commented on the manuscript at different stages. The book would be very different, and much the worse, were it not for the assistance of these people. Although they tried hard to remove all of our errors, they did not succeed in every case, and for this you should blame only us. In no particular order, these people are Brendan Bunting (University of Ulster, Northern Ireland), Mark Mugglestone (University of Derby), Germa Coenders (University of Girona, Spain), Dianne Phillips (Manchester Metropolitan University), Rick Hoyle (University of Kentucky, USA), Diane Miles, Phil Banyard (Nottingham Trent University), Susanne Hempel (Halle University, Germany), Fiona Bailey (Deakin University, Australia), Bill Anderson (University of Ulster at Magee), Viv Brunsden (Nottingham Trent University), David French (Cambridge University) and one anonymous reviewer.

More information about the book, including datasets and updates, can be found at http://www.jeremymiles.co.uk/regressionbook. If you have any comments or questions at all, we would be very interested to hear from you. We can be contacted via the Web page listed, or at regression@jeremymiles.co.uk.

PART I

I NEED TO DO REGRESSION ANALYSIS TOMORROW

1 Building models with regression and correlation

1.1 What are models?

How tall do you (the reader) think we (the authors) are? Decide your answer before you read on.

Your answer may have been, 'What a stupid question' or 'How should I know?' If you think about it, the question is not *very* stupid, and you should have some idea. You can see (from our names on the front cover) that we are both male. You can safely assume that we are not so young that we are not fully grown, and we will tell you that we did not start smoking until we had almost reached our full adult heights, thus stunting our growth only a little. Given this, you know that we are not 4 inches tall, nor are we 100 feet tall. A reasonable guess, given the information that you have, would be that we are of about average height for males – approximately 5 feet and 10 inches (1.77 m). We are in fact both slightly over 6 feet (1.83 m) tall, so if you had guessed the average height, your guess would have been a couple of inches out – not bad really.

What you did was to build a model. Your model was 'The authors are 5 feet 10 inches tall.' This is a model of our heights – it is a very simple model, but a model nonetheless. A model is a representation of the world, but it is rarely a perfect representation. There will always be some differences between the model and the world, that is some error. If we could build a perfect copy, it would not be a model, it would be a duplicate.

What you did when trying to guess our height was to pick a value that had as little error in it as possible. The value with the least chance of error would be the average height. There is more chance of us being average than being unusual – that's what averages are. Your model was as close as possible to the data (our heights), but there was some error remaining. A different way of saying this is that:

2 APPLYING REGRESSION AND CORRELATION

$$DATA = MODEL + ERROR$$

This is a very important statement, which we shall refer to a lot throughout this chapter, so let's look at a practical example.

Imagine that we have some data on the number of books on research methods and statistics that a small group of psychology students has read during their studies. Table 1.1 shows these data.

TABLE 1.1

Name	Number of books read
Anne	2
Bob	4
Carol	1
David	0
Esther	3

If we want to model those data, we could do it by repeating the numbers. We could say 2, 4, 1, 0, 3. The numbers that make up the model are known as *parameters*, and this model has five of them.

In terms of:

$$DATA = MODEL + ERROR$$

DATA are equal to MODEL, and so ERROR is zero; there is no difference between the model and the data. This model is a perfect representation of the data. However, there is a problem with this model. There were five numbers in the data, and there are five numbers (or parameters) in the model, so the model has not summarised anything — it is not really a model, but a duplicate. A model should be a simple, or parsimonious, representation of a phenomenon. In this case the model *is* the data, and we are back where we started (so in fact we could argue that this is not a model at all). That the model and the data are exactly the same is not much of a problem when we are dealing with five numbers, but if we are trying to summarise 500 numbers, this approach will not work, so we need a different approach.

If we want to model the data with one parameter, we could use the *mean*. This is what people commonly call the *average* (although it is better to avoid this term as it has more than one definition). To find the mean score, we add together the five numbers (sum the set of numbers), and divide this sum or total by the number of people:

$$2 + 4 + 1 + 0 + 3 = 10$$
$$10/5 = 2$$

We have calculated that the mean number of books on statistics read by these psychology students is two.[1] We have used the mean as a simple model, which has one parameter, and describes the data.

1.2 Least squares models

1.2.1 A very simple model

We saw in the previous section that we used a model that contained one parameter, the mean, to represent a set of five numbers. We used the mean for a good reason because it summarised the data: with just one number, it gives us a general idea about a whole set of numbers.

The mean is a special type of model; it is a *least squares model*. In this section, we shall see what we mean by a least squares model, and find out why the fact that the mean is a least squares model is important. Remember that (we said we would keep coming back to this):

$$\text{DATA} = \text{MODEL} + \text{ERROR}$$

DATA was the number of books read by each student; the numbers 2, 4, 1, 0, 3. Our model is the mean, the number 2. The difference between the model and the data is ERROR. If:

$$\text{DATA} = \text{MODEL} + \text{ERROR}$$

it is also true that:

$$\text{ERROR} = \text{DATA} - \text{MODEL}$$

Table 1.2 shows the number of books that each student has read along with the differences between the number for each student and the model (the mean). The difference between the model and the number of books read by a particular student is what we call the *error* for that student.

TABLE 1.2

Name	Number of books read (x)	Mean ($\bar{x}$)	Error ($x - \bar{x}$)
Anne	2	2	0
Bob	4	2	2
Carol	1	2	−1
David	0	2	−2
Esther	3	2	1

In statistics, the errors (or differences) such as those shown in the table are sometimes called *residuals*. The residuals are what are left over after the model (mean) has been taken away from each student's score. They are the difference between the score predicted by the model and the score that each individual actually has.

We said earlier that the model we were going to select was the model that gave the least error. We picked the mean height for males as your model for the heights of the authors of this book because that would have been your best guess. The score for each person is the mean plus (or minus) some error.

Often in statistics, we want to refer to a whole set of numbers, rather than just one individual number. For example, where we have a set of numbers that refer to the number of books read, we would call this x. We will often call this a variable, as it is something that can vary between people. You may also see it referred to as a 'vector' in more mathematically inclined texts. We can refer to the first number in the variables (or vector) as x_1, the second as x_2, etc. We can write Table 1.2 as:

$$x_1 = \bar{x} + e_1$$
$$x_2 = \bar{x} + e_2$$
$$x_3 = \bar{x} + e_3$$
$$x_4 = \bar{x} + e_4$$
$$x_5 = \bar{x} + e_5$$

The above list is the statistical way of saying that the score for the first person is equal to the mean plus the residual for the first person, the score for the second person is equal to the mean plus the residual for the second person, and so on. In statistics, we can use the subscript i for the individuals. Instead of writing out the full set of equations, as we did above, we can write:

$$x_i = \bar{x} + e_i$$

This means 'take this equation and repeat it for every person'. Sometimes statisticians are a little bit lazy, and do not bother writing i. Instead they write:

$$x = \bar{x} + e$$

We have been a little bit naughty here, and this can catch out the unwary. This time x and e do not refer to separate individual numbers but to the whole column (or row) of individual numbers, for instance the column of five numbers, which is shown in Table 1.2.

Now we shall calculate the total amount of error that is remaining when we use the mean as a model. We could see how much error we have got by adding together all the errors (or residuals) by reading these values off from the column titled 'Error' in Table 1.2:

$$\text{ERROR} = 0 + 2 + (-1) + (-2) + 1 = 0$$

When we total the errors, both positive and negative, the result in our model seems to be zero. However, we know that the error is not equal to zero; not everyone read two (the mean) books − whether positive or negative (too high or too low) an error is an error. The zero value is the result of negative errors cancelling out the positive errors. One way to get a result in which the negative errors do *not* have the effect of cancelling the positive errors is to get rid of minus signs by squaring the values of the residuals. Squaring gets rid of the minus signs because the square of a negative value is always a positive value. So we could square all of the values and calculate the amount of (squared) error *per person*.

If we add the squared error for each person and divide by the number of people, we get the mean squared error per person. (If we did not divide by the number of people, the error would increase when we had more people; if we had twice as many students, the error would be 20, when the model was just as good.) So, in mathematical terms, we are calculating error as:

$$\text{ERROR} = \frac{(0)^2 + (2)^2 + (-1)^2 + (-2)^2 + (1)^2}{5}$$

However, because we took the square of each of the numbers, it would be better if we took the square root:

$$\text{ERROR} = \sqrt{\frac{(0)^2 + (2)^2 + (-1)^2 + (-2)^2 + (1)^2}{5}}$$

$$= \sqrt{\frac{0 + 4 + 1 + 4 + 1}{5}}$$

$$= \sqrt{\frac{10}{5}}$$

$$= 1.41$$

When we refer to error from this point, we will usually be referring to the mean of the squared deviations from the mean.

You may recall from previous work that you have done that we have just calculated the sample *standard deviation*. This is a measure of dispersion of a set of numbers. The formula, put more simply, is:

$$\text{SD} = \sqrt{\frac{\sum(x - \bar{x})^2}{N}}$$

where x refers to each value, $\bar{x}$ is the mean, N is the number of people and the symbol Σ means 'take the sum of'. The standard deviation (SD) is a measure of error – it is a measure of how far the scores vary from the mean.

1.2.1.1 Why is the mean a least squares model?

We said earlier on that the mean was a least squares model, and now we will show why that is the case.

6 APPLYING REGRESSION AND CORRELATION

Let us see what would happen if we were to use some value other than the mean as the model to calculate the amount of error (using the standard deviation to represent error).

If we use the value 1 and carry out the same calculation (please forgive us for not showing all of the steps):

$$SD = \sqrt{\frac{(2-1)^2 + (4-1)^2 + (1-1)^2 + (0-1)^2 + (3-1)^2}{5}}$$

Similarly, if we use a value of 4:

$$SD = \sqrt{\frac{(2-4)^2 + (4-4)^2 + (1-4)^2 + (0-4)^2 + (3-4)^2}{5}}$$

$$SD = 2.45$$

Both of these values are higher than the error we found when we calculated the standard deviation using the mean.

The chart in Figure 1.1 shows how the value of the standard deviation (error) changes as the value we use for the mean changes. As you can see, the value for error reaches its lowest point when the value we use for the mean is 2.

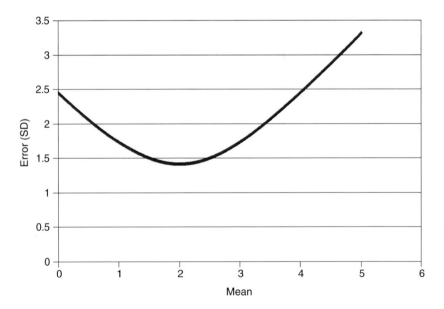

FIGURE 1.1 *The relationship between the mean and error*

Going back to:

$$\text{DATA} = \text{MODEL} + \text{ERROR}$$

if we consider ERROR to be the sum of the squared residuals from MODEL, the value of MODEL that gives the least ERROR is the mean. This reveals that the mean is one of a special class of models that are very important in statistics. It is a *least squares estimator*. It is the value for MODEL that gives the lowest value possible of ERROR, when ERROR is the sum of squared residuals. The mathematical proof is surprisingly simple, and is provided (along with lots more information) in Judd and McClelland (1989).

To conclude, the mean is a method of modelling (summarising) a set of numbers, which minimises the error (standard deviation) of that set of numbers.

1.2.2 The standard error of the mean

We now have a method for modelling a set of data, which is to use the mean. This mean describes our set of data. However, the data only tell us about our sample of individuals. Usually we are not interested in knowing things about the sample – we want to know about a much more general group, the *population* from which the sample was taken. The population is the larger group of people from whom the sample group was selected. When we have modelled our sample, we know the value of the parameters of the model for the sample, and these parameters provide an estimate of the model parameter for the population, *but it is only an estimate*. This model is called a sample estimate of a population parameter, or simply a *parameter estimate*.

We usually do not know the real parameter value for our population (if we did, we would not need to take a sample to estimate it), but we can make an educated guess about how close we are to the real answer by using a statistic called the *standard error of the mean*. If we took a large number of samples from the same population, each sample would be different, and so it is likely that the mean for each sample would be different. If we plotted these means on a histogram, we would find the *sampling distribution of the mean*, and we will usually find that this distribution has a specific shape, known as the normal distribution (we will encounter this distribution again in Chapter 4). Once we know the sampling distribution of the mean, we know the kind of values that we are likely to get if we took another sample and calculated the mean for that new sample.

If we only have one sample, we can *estimate* the standard deviation of the sampling distribution, using the standard error (we know that this is getting a bit tricky early on in the book, but bear with us, it will be over soon). When we have calculated the standard error, we can estimate how close we are likely to be to the true (population) mean.

We are going to need an example that has a bigger sample – we cannot learn much from a sample of five people (after all we might have accidentally sampled the five class geniuses, or the five laziest people in the class). The data in Table 1.3 are a sample of 40 students asked how many statistics books they have read.

TABLE 1.3

0	2	2	3
1	1	3	4
0	4	1	4
2	1	0	3
4	0	3	1
4	1	3	2
1	3	2	0
4	0	2	4
3	1	3	0
0	4	2	2

The mean of this dataset is 2.0, and the standard deviation is 1.43. If we want to know how near this mean value is to the mean value in the whole population we use the standard error.

The standard error of the mean of a set of numbers is given by:

$$\text{se}(\bar{x}) = \sqrt{\frac{\text{sd}^2}{n}}$$

where $\bar{x}$ refers to the mean, $\text{se}(\bar{x})$ is the standard error of the mean, sd is the standard deviation, and n is the number of people in the dataset. The standard error of the mean for the number of statistics books students have read is therefore:

$$\text{se}(\bar{x}) = \sqrt{\frac{1.43^2}{40}} = 0.23$$

This means that we know that if we calculated this mean and standard error, 68% of the time the true mean from the population that the sample was drawn from is between 1.77 (2 − 0.23) and 2.23 (2 + 0.23).

The standard error estimates how close we are to the true mean for the population. Because we can assume that the distribution of the means is normal, and we know the shape of the normal distribution (Figure 1.2), we know a great deal about what values are expected. We know that (approximately) 68% of the time, the true population mean will lie within one standard error of the mean, that approximately 95% of the time the population mean will lie within two² standard errors of the mean, and that approximately 99% of the time the population mean will lie within three standard errors of the mean.

However, knowing that 68% of the time the mean would be within a certain value is not very useful – a lot of the time (32%) we would be wrong. A much more useful value is 95%: knowing that, given our data, 95 times out of 100 we would know the true mean to be within a certain range. The measure of (approximately) two standard errors either side of the mean is often called the

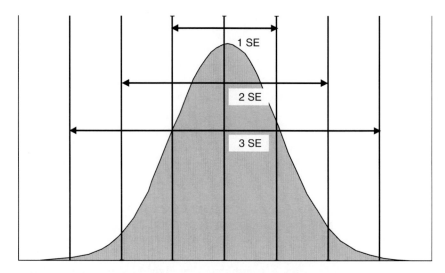

FIGURE 1.2

95% confidence interval (95% CI), because we are 95% confident that this interval includes the population value. The upper 95% CI is given by the mean + 1.96 SEs and the lower 95% CI is given by the mean − 1.96 SEs. The CIs are given by:

$$CI = \bar{x} \pm 1.96 \times SE$$

$$\text{Upper CI} = 2 + (1.96 \times 0.23) = 2.45$$

$$\text{Lower CI} = 2 - (1.96 \times 0.23) = 1.55$$

Thus, if we say that the mean number of books read by students in the next sample taken from that population from which we took our sample is between 1.45 and 2.55, we will only have a 5% chance of being wrong.

Throughout this book, we shall try to avoid complex calculations, preferring to concentrate on the principle behind the calculations. Before you utter sighs of relief, be aware that having to learn principles will make your job harder because you will have to *think* more. We could have shown you how to calculate the mean and standard deviation with two equations; instead we showed you the reasoning behind the calculations for the mean and standard deviation.

1.3 Modelling relationships

In many applications simply estimating a population parameter from a sample estimate is sufficient. When we do this, we have a simple model, which contains one parameter.

10 APPLYING REGRESSION AND CORRELATION

Regression analysis is a technique for modelling the relationships between two (or more) variables. Often in psychology, simply estimating a one-parameter model is not sufficient; we want to estimate the relationship between two or more variables.[3] For example, do students who have read more statistics books achieve higher marks in their statistics courses? One variable is the marks achieved by students, the other is number of statistics books read.

Table 1.4 shows the data from the same group of students that we looked at earlier, but now we have the final marks that they have achieved (in percentage terms). In the table, the students have been assigned numbers from 1 to 40. What we would like to know is: do students who read more books get better grades?

What sort of model are we going to need to work this out?

One way to start working on this problem is to draw a graph. Figure 1.3 is a scatterplot showing the number of books read by each of the students along the *x*-axis and the mark achieved by that student along the *y*-axis.

This chart seems to show a trend that the students who have read more books seem to be getting higher marks, but it is not clear. We want to know how many more marks a student is likely to get, if they were to read another textbook. In other words, we want to estimate the parameter that will be the *mean mark increase per book read*.

If we think about what that would look like, it would be a straight line; it may look a little like Figure 1.4.

If, like us, you wondered what was the point of studying maths at school, you may find the following something of a revelation. If you do not remember, that is OK.

Many people spent many hours in maths classes at school drawing straight lines based on equations. These equations usually looked something like:

$$y = mx + c$$

where x and y are the two variables, which are represented on the two axes, m is the slope of the line, and c is the y-intercept – the value of y when x is equal to zero (or the point where the line crosses the y-axis). The equation is a representation of a straight line on a graph.

If $m = 1$ and $c = 2$, then we say that:

$$y = (1 \times x) + 2$$

There is only one graph we can draw from this equation (Figure 1.5). When x is 0, y is 2; when x is 1, y is 3; etc. With any straight line graph, if we know the value for x, we can read off the value for y.

When we only wanted to represent one set of data using our model, one parameter, the mean, was fine. Now we would like our model to represent a (linear) relationship between two variables, and to do this we will need two parameters. When we know those two parameters, we will know what the line looks like for our data, where x is the number of books, and y is the grade.

TABLE 1.4 Dataset 1.1

Number	Books	Marks (%)
1	3	56
2	1	57
3	0	45
4	2	51
5	4	65
6	4	88
7	1	44
8	4	87
9	3	89
10	0	59
11	2	66
12	1	65
13	4	56
14	1	47
15	0	66
16	1	41
17	0	56
18	0	37
19	1	45
20	4	58
21	4	47
22	0	64
23	2	97
24	3	55
25	1	51
26	0	61
27	3	69
28	3	79
29	2	71
30	2	62
31	3	87
32	2	54
33	2	43
34	3	92
35	4	83
36	4	94
37	3	60
38	1	56
39	2	88
40	0	62

If you went to a similar school to us you may have had to draw 'lines of best fit' onto scattergraphs. You took a ruler, put it on the graph until it looked like it was in the right place and then drew a line. We want to do a similar procedure — to find the line of best fit — but we want to do it in a slightly more scientific fashion. In fact what we would like is the *least squares estimate* of the line. To do this we need to devise a model that has two parameters, m and c. Finding this line is the basis of what regression analysis is about. In regression analysis the c remains as a c (although sometimes you

12 APPLYING REGRESSION AND CORRELATION

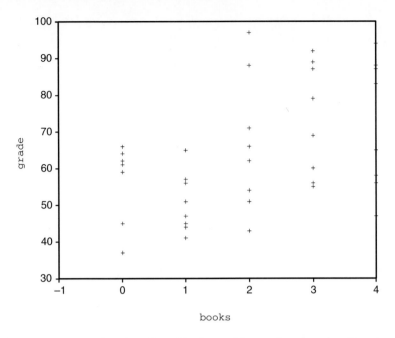

FIGURE 1.3 *Scatterplot of number of books read* (books) *and grade achieved* (grade)

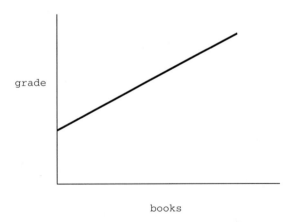

FIGURE 1.4 *Representation of the mean increase in marks per book read*

may see it referred to as b_0), and is often referred to as the *y*-intercept (because it is the value where the line intercepts the *y*-axis), and the slope, *m*, is called *b*, sometimes referred to by the Greek letter β (beta), or as a regression line. Additionally, because our model is not perfect, and is likely to contain some error, we include the symbol *e* for the error in the equation. The equation looks like this:

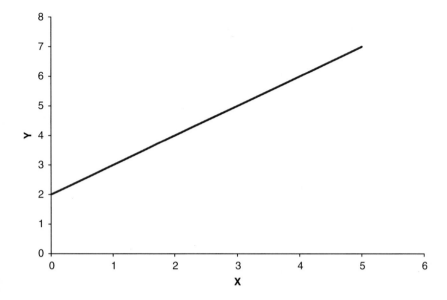

FIGURE 1.5 *Graph of* y = (1 × x) + 2

$$y = bx + c + e$$

This equation says that the values for *y* are equal to the value for *x*, multiplied by some value (*b*), plus a constant (*c*). In addition, because our model is not perfect, the values for *y* will deviate from their predicted value by a random amount (*e*). The error term (*e*) is not necessary in the calculation and interpretation of this type of regression equation, so for simplicity we will usually drop the term from the equation in future.

To show how the regression line is calculated, we will start with a simpler example. Let us go back to the type of model that only had one parameter, the mean, which we calculated from the dataset contained in Table 1.4. To make life easier we will consider only the first five participants (reproduced in Table 1.5). The graph in Figure 1.6 shows a scatterplot of these participants with the number of books read on the *x*-axis and the grades achieved on the *y*-axis. The one-parameter model (that was the mean) that we selected was the model that gave us the lowest value for error when error was the sum of squared residuals. The mean score (which is the predicted score) of 54.8 for the grades is marked on the graph, and the residuals are calculated and labelled on the graph.

The error (or standard deviation) is equal to:

$$\sqrt{\frac{1.44 + 4.84 + 96.04 + 14.44 + 104.4}{5}} = \sqrt{44.232} = 6.65$$

14 APPLYING REGRESSION AND CORRELATION

TABLE 1.5

Number	Books	Marks (%)	Predicted score (mean)	Residual	Residual squared
1	3	56	54.8	1.2	1.44
2	1	57	54.8	2.2	4.84
3	0	45	54.8	−9.8	96.04
4	2	51	54.8	−3.8	14.44
5	4	65	54.8	10.2	104.4

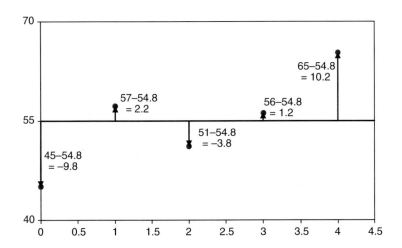

FIGURE 1.6

To draw a sloping line, we need two parameters, one to tell us where to start the line and one to tell us how steep to make the line. We want to draw the line that minimises error (the sum of the squared residuals). Because we can describe a line using only two parameters, we will have a model that has two parameters.

The model that we want is (again) the one that minimises error. We need to choose values for b (the slope) and c (the constant) which minimise error. With the small dataset, these two values are $b = 3.9$, $c = 47$. When we know these two values, we can calculate the values that would be predicted by this model, by using the formula:

$$\text{Predicted value} = 3.9 \times \text{books} + 47$$

We can replace the previous values of predicted score, residual and residual squared that were shown in Table 1.5 (when we used the mean as a model) with the values from our new model, which has two parameters. These values are shown in Table 1.6.

We can recalculate the amount of error in this model in just the same way as we calculated the amount of error with the mean. Error is calculated using:

TABLE 1.6

Number	Books	Marks (%)	Predicted score	Residual	Residual squared
1	3	56	58.70	−2.70	7.29
2	1	57	50.90	6.10	37.21
3	0	45	47.00	−2.00	4.00
4	2	51	54.80	−3.80	14.44
5	4	65	62.60	2.40	5.76

$$\sqrt{\frac{7.29 + 37.21 + 4.00 + 14.44 + 5.76}{5}} = \sqrt{13.74} = 3.71$$

This error value, of 3.71, is considerably lower than the error that we had with the mean.

We can also redraw the graph with the new model represented by the sloping line, and the old model (the mean) still represented by a flat line (Figure 1.7). The new residuals are also shown, and it can be seen that these are smaller than the residuals with the old model. This line represents the lowest value for error that we can find using a straight line, and hence it is a least squares model, in just the same way that a mean was a least squares model. We have taken the mean (which we represented as a flat line) and turned it to make it into a sloping line, not to give the mean score for everyone, but to find the mean score per book read.

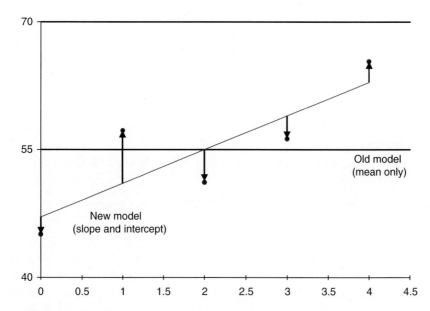

FIGURE 1.7

The equations to calculate the parameters of this model (i.e. the position of the line) are a touch fiddly, but we do not need to know them because we have a computer that will work out our calculations. (See Appendix 1 for details of the equations, and Appendix 2 for details of how to carry out a regression analysis in a range of programs.)

In the example above, we analysed only a small subset of the data because we wanted you to be able to see what was going on. Now we will analyse the whole dataset using a computer program. Most computer programs produce a lot of output, but the only part of the output we are interested in (for now) is the part that tells us what the values are for the unstandardised regression coefficient (b) and the constant (c). These values are shown in Table 1.7. The constant, c, has the value 52.08, and the slope parameter, b, has the value 5.74.

TABLE 1.7

	b	Std error of b	Standardised b	t	Sig.
(Constant)	52.08	4.035		12.91	<0.001
books	5.74	1.647	0.492	3.48	0.001

We go back to our straight line. We can model it as:

$$y = bx + c$$

We can then substitute the values given to us by our computer program into the equation to form the model that minimises error. The model that minimises error for these variables is therefore:

$$\text{grade} = 5.74 \times \text{books} + 52.1$$

In other words, the average student who reads no books will achieve a mark of 52.08 in the exam, and each additional book they read will increase their grade by 5.74 percentage points. (Please note: there are a large number of qualifications required for this statement, which will be covered in the following chapters.)

We can now redraw the scattergraph and add the line of best fit (Figure 1.8).

1.3.1 The standard error and significance of parameter estimates

In Table 1.7 we had an estimate of the slope coefficient and the constant. We know that the program could calculate a standard error of the mean when the mean was our parameter. Similarly, the program can also give a standard error of the slope coefficient and the constant, our new parameter estimates. Again, we can use standard errors to calculate the 95% confidence intervals (CIs) of these parameters as we did with the mean. The upper and lower CIs indicate the range of values in which we are likely to find the true (population) value. The calculation is the same as before. For b (the slope):

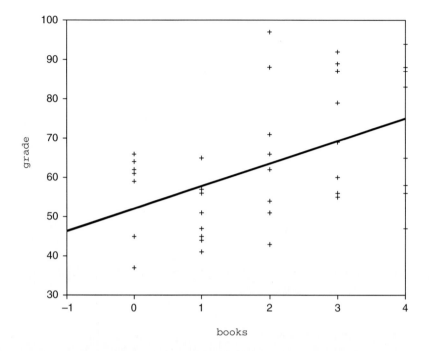

FIGURE 1.8 *Scatterplot of* books *and* grade, *showing line of best fit*

$$CI = b \pm 1.96 \times SE$$
$$\text{Lower CI} = 5.74 - 1.96 \times 1.647 = 2.512$$
$$\text{Upper CI} = 5.74 + 1.96 \times 1.647 = 8.97$$

Thus we can be fairly sure (95% sure) that repeated samples from the same population would yield values between 3.22 and 8.97 for b.

And for c (the constant):

$$CI = c \pm 1.96 \times SE$$
$$\text{Lower CI} = 52.08 - 1.96 \times 4.035 = 44.17$$
$$\text{Upper CI} = 52.08 + 1.96 \times 4.035 = 59.99$$

Thus we can be fairly sure that the constant will lie between 44.17 and 59.99.

The standard error of the slope can also be used for something more useful, that is for determining whether the figure given by the slope is statistically significant. The big question we would like to know the answer to is: 'Does reading more books *really* have any effect on grades?' We have found that *in our sample*, there is a positive relationship between number of books read and grade, but what we want to know is whether this is the case for *the population*. Is it possible, or likely, that the population values for each of those parameters is equal to zero? In psychology, we are not usually interested in the intercept, so we will ignore that figure for now. We are interested in the slope (*b*) because it indicates the amount of change in the dependent variable (grade) that we would expect for a change of one unit in the independent variable (books read). It is most important to know whether the value we have for the slope is *statistically significant*. Essentially, if the value of the slope in the sample is statistically significant, then the value of the slope in the whole population can be considered to be different from zero. In Table 1.7 a *t*-value is reported for the slope (3.48), which has an associated probability of 0.001. This is a statistically significant result at the 5% level (or indeed the 1% level; see Mohr (1990) for a good introduction to significance testing). More formally we can say that the null hypothesis of no linear relationship between the variables can be rejected. This result means that if the null hypothesis were true (and there is in fact no effect), we would expect a correlation of this magnitude to be found less than one time in a thousand.[4] It should be remembered that this is all that such significance tests mean; a small *p*-value does not mean that the slope is 'very significant', just that it is likely to be greater than zero in the population.

1.3.2 Standardised estimates

One problem with the estimate of the slope of the line is that it is dependent upon the measurement scale that is used. Comparing results from the hypothetical study above with the results of similar studies is only possible if the other studies used the same scale. We found an increase of approximately 6 percentage points per book read. What would happen if someone else found an increase of 2 marks out of 10 per book chapter read? Another study may find an increase of 0.01 grades (on a scale from A to F) per page of book read. We clearly would have difficulty comparing these results. One way round this is to ask everyone to use the same scale — we could ask them to turn chapters into books, and marks into percentages.

A different way of making everyone use the same scale is to standardise the measures that are used. A standardised score has a mean of 0 and a standard deviation of 1. These scores are sometimes referred to as *z-scores*. To convert a value into a standardised score, or *z*-score, we subtract the mean and divide by the standard deviation. Put another way:

$$z_i = \frac{x_i - \bar{x}}{\text{sd}}$$

Alternatively, many statistics packages can create standardised scores automatically. A scattergraph of the standardised variables is shown in Figure 1.9.

When we calculate the line of best fit using standardised variables, the intercept will always be zero. We can therefore remove this parameter from our model thus making our model simpler.

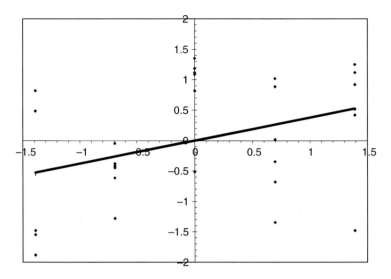

FIGURE 1.9 *Scattergraph of z-scores of* books *and* grade

What we are interested in is the b parameter (i.e. the slope of the line, the increase in y as x increases by one unit). It appears from Figure 1.9 that an increase of one unit along the books (x-)axis is associated with an increase of almost 0.5 on the grade (y-)axis. If we did the regression analysis, we would find that the b-value was actually 0.492. If we look at the original table produced by the first regression analysis (Table 1.7), we find that the value 0.492 already appears in the column of standardised coefficients — it is usually called the standardised slope, or standardised beta. Because the standardised slope is rescaled, it does not matter what the original variable was — whether it was books, chapters, pages or words, the standardised slope will always be the same. The standardised slope therefore gives us a value that we can compare, and one that is meaningful across studies that have used a different measurement scale.

An important warning

It is OK to compare beta values across studies if you have only one independent variable. In the following chapter(s) we will consider cases where there is more than one independent variable, and in such cases it is (usually) inappropriate to compare beta values across studies.

Where there is a single independent variable, the standardised b (beta) coefficient is equal to the Pearson correlation coefficient — with which you may already be familiar. Actually, it is more than equal to: it is the Pearson correlation (see Appendix 1 for the proof), or alternatively you can confirm

20 APPLYING REGRESSION AND CORRELATION

the truth of this statement by calculating the Pearson correlation using a statistics package.

We think that this demonstrates something that we will be coming back to regularly throughout this book. Some students believe that statistics is about a series of unrelated equations that are used in specific situations, and that each test must be learned anew and afresh, and similarly that each procedure in a statistical package looks a bit different and must be learned from new. In fact, almost all statistical tests are based on very similar principles and are very closely related to one another. We have arrived at the correlation between two variables in three different ways: through regression; through standardising the variables and producing a graph; and, finally, through a correlation procedure. In later chapters, we will look at how other statistical tests can be carried out using regression. For now, we will look a little more at correlations.

1.4 Looking more at correlations

1.4.1 Correlations and scattergraphs

A correlation is a measure of the extent to which two variables are linearly related. This relationship remains a linear relationship regardless of their measurement scales. The correlation is given the symbol 'r'. A correlation is just a number that represents a special case of a regression line in which the intercept is zero and the variables have a mean of 0 and a standard deviation of 1. The correlation can be thought of as the extent to which the scattergraph of the relationship between two variables fits a straight line. If the points all fall on a straight line going from bottom left to top right the correlation will be equal to +1.00, a perfect linear association. If the points all fall on a straight line going from bottom right to top left the correlation will be −1.00. Figure 1.10 shows a scattergraph representing a correlation of 1.00; Figure 1.11 shows a scattergraph representing a correlation of −1.00.

When you have used correlations more you will become more used to interpreting scattergraphs, but here are a few more to help get you used to the idea of how the correlation coefficient (r) and the scattergraph relate to each other.

In the chart in Figure 1.12, the points are all close to a line that we could draw through the diagonal, and so the correlation is very near to 1.00. In fact, it is 0.95.

It can be seen in the scattergraphs in Figure 1.13 that when the points lie closer to the shape of a line drawn in the diagonal the correlation is higher. The sign of the correlation gives the direction of the diagonal line.

1.4.2 Correlations and variance

The variance of a variable is the square of the standard deviation (you might remember that when we calculated the standard deviation the last stage

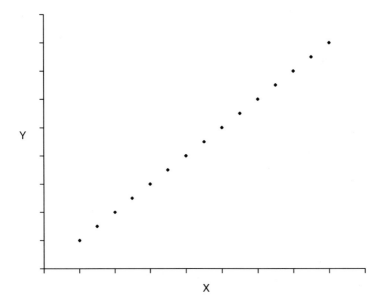

FIGURE 1.10 *Scatterplot showing* r = 1.00

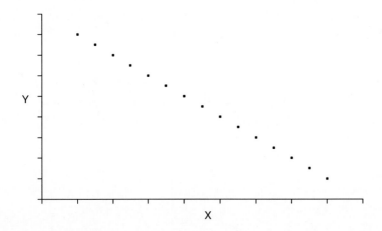

FIGURE 1.11 *Scatterplot showing* r = −1.00

was to take the square root — to calculate the variance you just miss the last stage). As the name may suggest, the variance is a measure of the amount of variability of a variable, which is how much the scores deviate from the mean.

We will return to our previous example of books read and grades. The variance of the grade is equal to the amount of *error* in grade. Figure 1.14 shows the scattergraph of grade and books with the mean score for the grades represented as the central horizontal line. The total variance is the average size of the squared deviations from that line. This value we can

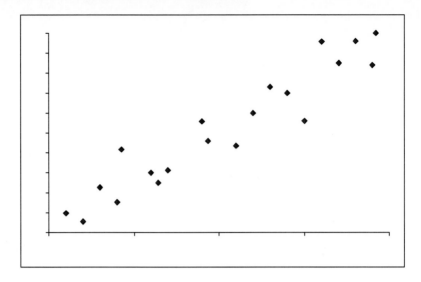

FIGURE 1.12 r = 0.95

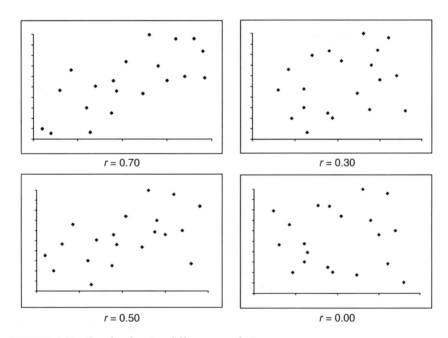

FIGURE 1.13 *Graphs showing different correlations*

calculate by hand, or by using a statistics package. It is found to be 279.1. (You might be familiar with the total of the squared deviations from ANOVA, where it is called the total sum of squares, or sometimes SS_{total}.)

Now we can calculate (by finding b and c) the position of the regression line, which is also known as the least squares line of best fit. We have

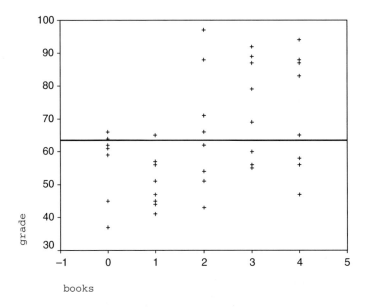

FIGURE 1.14 *Scattergraph with line showing the mean of* grade

calculated the position of this line before but we will do it again here. This line is the one shown in the graph of Figure 1.15. We can work out the error again, by calculating the average squared deviation from this line.

To calculate the amount of error we find the predicted value for every point on the line using the regression equation. The equation that gave the best estimate of the grade a person would get based on the number of books they had read was:

$$\text{grade} = 5.74 \times \text{books} + 52.08$$

Now we can take the previous table of results, and calculate the expected values and the deviation from the expected value.

We have shown the first person in the dataset on the graph by representing them as a large solid circle. For this person, the grade predicted by our model is given by:

Predicted grade = 5.74 [the slope] × 0 [number of books read]
     + 52.08 [the intercept]

     = 52.08

The person's actual mark is 45 and therefore the residual, the difference between the predicted value and the actual value, is −7.08 (45 − 52.08). This value represents the deviation from our model, where our model is given by the two parameters b and c (i.e. our model is the line of best fit). We could

24 APPLYING REGRESSION AND CORRELATION

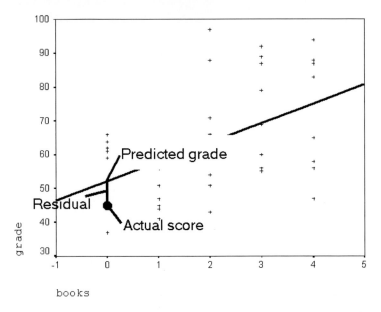

FIGURE 1.15 *Scatterplot of* books *and* grade, *showing the line of best fit*

continue to do this for each person, as shown in Table 1.8, and then we could work out the variance of the residual scores by finding the mean of their squared values.

TABLE 1.8 *Predicted score and residuals for first eight cases in dataset 1.1*

N	Books	Grade	Predicted grade	Residual
1	0	45	52.1	−7.1
2	1	57	57.8	−0.8
3	0	45	52.1	−7.1
4	2	51	63.6	−12.5
5	4	65	75.0	−10.0
6	4	88	75.0	13.0
7	1	44	57.8	−13.8
8	4	87	75.0	12.0

Note that we have rounded these values to one decimal place.

Most statistics packages will calculate and save both the predicted grades and the residual scores (see Appendix 1) as shown in Table 1.9.

TABLE 1.9 *Descriptive statistics for residuals and predicted scores*

	Mean	Std deviation	Variance
Predicted scores	63.5	8.22	67.53
Residual scores	0.0	14.55	211.59

The first thing to note is that the variance of grades is equal to the variance of the predicted grades plus the variance of the residuals (67.53 + 211.59 = 279.12).[5] This means that the variance of the residuals represents variance in grade unaccounted for by the model, whereas the variance of the predicted grades represents variance in grade accounted for by the model. If we divide the variance of the predicted scores by the variance of grade, we will find the proportion of the variance in grade that has been accounted for by books:

$$\frac{67.526}{279.074} = 0.2419$$

Finally, we take the square root of this value:

$$\sqrt{0.2419} = 0.492$$

If you have been very observant and concentrated hard you may have noticed that this value is equal to the value we found for the correlation that we calculated beforehand between the two variables. This result is important for the interpretation of correlation coefficients. It means that if you square a correlation, the resulting figure is equal to *the proportion of variance that the two variables share*.

1.4.3 Correlations and size

How large is a large correlation? How large a correlation is large enough? No easy answers to this question exist, as it all depends on what you are doing, and what you expect to find. However, when considering a correlation it is always useful to determine if it is statistically significant. Remember that a significance test for the estimate of the slope told us if the slope was likely to be zero or not. Also remember that a correlation is only another way of expressing the slope. It should be clear then that the significance test associated with a correlation tells us whether the correlation is likely to be zero. A probability value for each correlation can be found (you can use the tables shown in Appendix 3 for this, but most statistics packages will calculate it automatically). If the probability value associated with the correlation is less than 0.05 (or whatever the chosen significance level) you can conclude that the correlation is unlikely to be zero. As with significance tests for the regression slope, the probability value for a correlation only indicates that the correlation is greater or less than zero – it should never be interpreted as an indication of the strength of the association. This is because a small correlation, for example 0.1, will be statistically significant (at the 0.05 level) with a sample size of 400 even though a correlation of this size is unlikely ever to be of any practical significance.

For describing the magnitude or strength of the association between variables, some guidance has been provided by Cohen (1988) who has defined a small correlation as having an absolute value of approximately 0.1, a medium correlation as 0.3 and a large correlation as 0.5 or greater.

Notes

1 You might be tempted to say, 'The average student has read 2 books on statistics.' We think that this is a dangerous path to go down, because you may find yourself making nonsensical statements like 'The average family has 1.7 children.' Clearly, no family has 1.7 children, so you are talking about something (the average family) that does not exist.

2 The actual value is 1.96 standard errors.

3 In many applications, a one-parameter model is sufficient. A health economist may be interested in estimating the number of people who are hospitalised with influenza, while a politician may be interested in assessing the number of people who are unemployed.

4 There is some debate amongst statisticians regarding what is actually meant by a p-value. If you would rather accept our very brief explanation, please do so, but if you would like to explore this further then see Gigerenzer (1993) or Pollard (1993). We should perhaps point out that the two authors of this book do not agree with each other regarding the precise meaning of a probability value.

5 You may be familiar with the between-groups, within-groups and total sums of squares, from analysis of variance. These values are equivalent.

Further reading

For further details about how regression analysis minimises the sum of the squared deviations from the mean, see Judd and McClelland (1989) *Data analysis: a model comparison approach*, Chapter 1. Cohen and Cohen (1983) *Applied multiple regression/correlation analysis for the behavioral sciences*, Chapter 1, contains more information on how the correlation is calculated, and Allison (1999) *Multiple regression: a primer* is a nice introduction to many aspects of regression analysis with a question and answer format throughout.

2 More than one independent variable — multiple regression

2.1 Introduction: multiple regression in theory

In the previous chapter, we examined the situation where we had one independent variable (number of books read) and one dependent variable (final grades for the course). In this chapter we will take this basic model and expand it to the case where we have more than one independent variable. The situation where we have more than one independent variable is usually known as *multiple regression*.

This chapter has two parts. In the first part, we will describe the ideas and concepts behind multiple regression. For simplicity, we will look first at the case where there are only two independent variables so we can give an overview of how multiple regression works. Then we shall look at a multiple regression analysis in a practical context.

2.2 What's multiple regression all about?

It may not surprise you to find out that it is all about:

$$\text{DATA} = \text{MODEL} + \text{ERROR}$$

In the previous chapter we developed a model for predicting a student's grades. We predicted the grades on the basis of the number of books the student had read. We expressed that model in terms of a regression equation:

$$y = bx + c$$

where y was the grade the student achieved, x was the number of books that the student read, b represented the increase in y (grade) per book read (x), and c represented the constant (sometimes called the intercept) the predicted grade of a student who read no books.

We also expressed that model in terms of a standardised regression equation, which was equal to the correlation:

$$r = 0.492$$

28 APPLYING REGRESSION AND CORRELATION

When people make attributions about the causes of events, they tend to invoke one specific cause. When asked, 'Why did you fail the research methods course?' students (in our experience) reply with statements like:

- 'I didn't like the lecturer, we just didn't get on.'
- 'The textbook was no good.'
- 'The exam was not fair.'

The same principle is true of most events that people make causal attributions about. They tend to attribute an effect to one single cause. However, the fact is that several events may have contributed to the result. For instance, the lecturer may not have liked you, the exam might not have been fair *and* you may not have read enough books. The fact that people tend to think of one factor as being the *cause* of an event we refer to as *the myth of monocausality*. The teacher, the textbook, the time of the lectures (too early on a Monday morning?), the hours of work you did, your natural ability, the type of work you did, the questions that came up, and of course luck, to name a few, were all factors in how well you did in the exam. If our model says simply that the final mark you get is due to the number of books that you read (and error), then our model is too simplistic. What we need to do is make a more complex model; that is, measure some other variables than just the number of books read that may have contributed towards your final grades. If we knew about these other factors, we would be able to make a more accurate prediction of your final grade than we can simply on the basis of the number of books you read. If we have two (or more) independent variables, we cannot use the regression equation given at the start of this chapter because it only contains one slope (b) and one independent variable (x). What we have to do is include a second slope and another independent variable. This new equation is:

$$y = b_1 x_1 + b_2 x_2 + c$$

Note that we have introduced subscripts. These simply denote the different slopes associated with each independent variable. So b_1 is the slope for the first independent variable x_1, b_2 is the slope for the second independent variable x_2, and so on. We will now use this new regression equation to extend the analysis presented in Chapter 1. The data presented in Table 2.1 show the same set of students that we used in the previous chapter but now we have added data that tell us how many statistics lectures they attended. This is in the column labelled `attend`.

We ran another regression analysis (as we did in Chapter 1), but used `attend` instead of `books` as the independent variable. We got the results shown in Table 2.2.

As in Chapter 1 we can turn this into a regression equation and say that:

$$\text{grade} = \text{attend} \times 1.883 + 36.998$$

TABLE 2.1 *Dataset 2.1*

N	books	attend	grade
1	0	9	45
2	1	15	57
3	0	10	45
4	2	16	51
5	4	10	65
6	4	20	88
7	1	11	44
8	4	20	87
9	3	15	89
10	0	15	59
11	2	8	66
12	1	13	65
13	4	18	56
14	1	10	47
15	0	8	66
16	1	10	41
17	3	16	56
18	0	11	37
19	1	19	45
20	4	12	58
21	4	11	47
22	0	19	64
23	2	15	97
24	3	15	55
25	1	20	51
26	0	6	61
27	3	15	69
28	3	19	79
29	2	14	71
30	2	13	62
31	3	17	87
32	2	20	54
33	2	11	43
34	3	20	92
35	4	20	83
36	4	20	94
37	3	9	60
38	1	8	56
39	2	16	88
40	0	10	62

TABLE 2.2 *Regression analysis, using* grade *as the dependent variable and* attend *as the independent variable*

	Slope (b)	Std error of slope	Standardised slope (beta)	t	Sig.
Constant	36.998	8.169		4.529	<0.001
attend	1.883	0.555	0.482	3.393	0.002

Each additional lecture attended increases a student's grade by 1.9%. We can also see that the standardised regression coefficient (i.e. the correlation) between the two variables is 0.482. This correlation of 0.482 is lower than the standardised regression coefficient between books and grade (which was 0.492), and so we might say that attendance is a better predictor of final grade in a course than is the number of books read (although this difference is very small, and not significant). We can also see from the significance value in the final column that there is a significant relationship – it is unlikely to have occurred if there was no relationship in the population.

We showed in Chapter 1 that the squared correlation gives the proportion of variance that is shared by the two variables or, in other words, the squared correlation gives the percentage of variance in the dependent variable that is accounted for by the independent variable. We saw that books predicted $0.492^2 = 0.242$, which is equal to 24.2% of the variance in grades. We have now done the same analysis with attend as the independent variable and we find that $0.482^2 = 0.232$, which is equal to 23.2% of the variance in grades. Does this mean that we have predicted a total of $24.2 + 23.2 = 47.4\%$ of the total variance in grade?

The answer is 'No' because some of that variance that is shared between books and grade is also shared between attend and grade. If and only if the correlation between attend and grade is equal to zero will the total percentage of variance explained be equal to 47.4%. For that to happen we have to say (and show) that the number of books on research methods and statistics a student reads is unrelated to the number of lectures they attend (which seems unlikely – more enthusiastic students attend more lectures *and* read more books). Therefore, we need to know a third correlation, the correlation between books and attend. We have calculated the correlation between all three variables and present all these variables in the correlation matrix shown in Table 2.3.

TABLE 2.3 *Correlation matrix of* books, attend *and* grade

	books	attend	grade
books	1.000	0.444	0.492
attend	0.444	1.000	0.482
grade	0.492	0.482	1.000

This reveals that the correlation between books and attend is equal to 0.444. The fact that these two measures correlate should not surprise us as more enthusiastic students are likely to attend more lectures and read more books.

What we are interested in is the correlation between books and grade, with the influence of attend removed or 'controlled for'. Similarly, we want the correlation between attend and grade with the influence of books removed, or 'controlled for'.

What do we mean by 'controlled for'? In experimental psychology, variables that we are not interested in are controlled using standardised conditions. In laboratory conditions we control such things as lighting, temperature, noise levels, and other variables that may affect performance. In a real-life study such as the example we are using, we cannot control other variables through standardisation. We cannot say to every student that they must read exactly four books as part of their course. (Well, we could – let's face it, we do – but experience tells us that a large number will not do so.) Similarly, we cannot tell every student to ensure that they miss two, and only two, lectures on the course. So instead of standardisation to control extraneous variables we use *statistical* control. The multiple regression analysis tells us what the relationship would have been between books and grade, if everyone had attended the same number of lectures. Similarly, it tells us the relationship between attend and grade if everyone had read the same number of books.

When we have one independent variable (books) we explain a certain proportion of the variance in the dependent variable (grade). When we introduce a second independent variable (attend) we want to know how much variance that variable explains. However, to do this we have a problem: books and attend are likely to be correlated, and if they are, they will share some of their variance, so how can we find the total amount of variance explained in grade?

Let us consider some hypothetical possibilities. The first possibility is that books and attend are uncorrelated. If this is the case we can safely add up the value for the proportion of variance accounted for by books and the proportion of variance accounted for by attend, and add them together to get a total variance accounted for in grade.

We have already seen that this is not the case for our data as the correlation between books and attend is not equal to zero – in fact we would be surprised if it turned out that there was no relationship between the number of books students read and the number of lectures they attended. So what we would really like to know is what proportion of variance does each of books and attend explain in grade, when the other is controlled for? That is, if everyone went to the same number of lectures what effect would reading more books have? In addition, if everyone read the same number of books, what effect would going to more lectures have?

Multiple regression calculates these proportions by taking into account the correlations between independent variables, and assessing the effect of each independent variable, when the other variables have been removed.

2.3 Multiple regression in practice

So if we enter both books and attend into the regression equation, we get estimates of the slope coefficients for each variable, controlling for the other variables. In the following section we will examine the output from a statistical

2.4 R and R square

The first part of the output we will consider is the model summary, shown in Table 2.4.

TABLE 2.4

R	R square	Adjusted R square
0.573	0.329	0.292

R is the multiple correlation, sometimes known as the *coefficient of determination*. R represents the total correlation between all the independent variables and the dependent variable. Recall from Chapter 1 that the value of R was the same as the value of the correlation between two variables. R squared, or R^2, is R which has been squared. We said in Chapter 1 that the square of a correlation was the same as a proportion of variance, and the same is true of R square — it represents the total amount of variance accounted for in the dependent variable by the independent variable(s). The value of R square can be interpreted as the *proportion of variance explained* by moving the decimal point two places to the right and expressing this value as a percentage. In this case we can conclude that books and attend explain 32.9% of the variance in grade.

When we did the regression analysis using books as the independent variable, we found that books accounted for 24.2% ($R^2 = 0.242$) of the variance. When we did the regression analysis using only attend as the independent variable, we found that attend accounted for 23.2% ($R^2 = 0.232$) of the variance in grade. If these two independent variables were not related to one another at all, we would find that together they would account for 23.2% + 24.2% = 47.4% of the variance in grade. But because attend and books are correlated with each other, the effect of them jointly is reduced to 32.9%.

2.5 Adjusted R square

Adjusted R square is a reduced value for R squared (R^2) which attempts to make an estimate of the value of R^2 in the population (rather than the sample). The reasoning behind this adjustment is that if another independent variable is added it is very unlikely that the correlation between that independent variable and the dependent variable will be exactly zero, even if it is zero in the population. It will almost always fluctuate around zero just because of sampling error. (Try it: you may know that a coin will on average come up heads 50% of the time, and tails 50% of the time, but throw it 50 times and see if it does — it is likely that you will not get 25 of each.) Because of this random variation from zero, R^2 will always go up a little when another independent variable is added. So adjusted R^2 is adjusted down to compensate for this increase in R^2. The larger the number of independent variables, the greater the downward adjustment in R^2 that will occur. In addition, the smaller the sample size, the greater the random variation from zero will be and therefore the larger the downward adjustment in R^2 that is required. The following shows the calculation of adjusted R^2:

$$\text{Adj. } R^2 = 1 - (1 - R^2)\frac{N-1}{N-k-1}$$

where N is the number of people and k is the number of independent variables. As N increases, the amount by which R^2 is adjusted downwards decreases, and as k increases, the amount by which R^2 is reduced increases.

2.6 Analysis of variance (ANOVA) table

The next part of the printout is the analysis of variance (ANOVA) table, reproduced in Table 2.5.

TABLE 2.5 ANOVA output from regression

	Sum of squares	df	Mean square	F	Sig.
Regression	2 633.513	1	2 633.513	12.130	0.001
Residual	8 250.387	38	217.115		
Total	10 883.900	39			

ANOVA in a multiple regression output may seem strange at first because it is generally used by psychologists to examine differences between group means. However, as the name implies, ANOVA examines variability and can be applied in this situation to look at the total amount of variance in the dependent variable, and how much of that variance is accounted for by the independent variables. The significance of the value F (called Sig. in the table) is the probability associated with R^2, that is the probability of getting a value of R^2 as high as it is *if* the actual value in the population is zero. This probability can be thought of as a significance value for the whole model or, equivalently, a significance value of R^2.

2.7 Coefficients

The final part of the printout is the coefficients. We met these before in Chapter 1, and they are shown in Table 2.6.

TABLE 2.6 *Parameter estimates*

	Slope (*b*)	Std error of slope	Standardised slope (beta)	*t*	Sig.
Constant	37.379	7.745		4.8	0.000
books	4.037	1.753	0.346	2.3	0.027
attend	1.283	0.587	0.329	2.2	0.035

As before in our regression equation, we have the constant, but whereas previously the constant was the expected value of the dependent variable (i.e. grade) when books was zero, now the constant is the expected value of grade when books *and* attend are zero. The unstandardised coefficients then tell us how much the student's reading of one more book or attending one more lecture should increase the student's grade. We find that reading one more book will increase a student's grade, on average, by 4.037 marks, and that attending one more lecture will increase a student's grade by 1.283 marks. But you might recall that in the previous chapter, we said that if a student read one more book, they would achieve a grade that was 5.7 marks higher. Why is the figure now only 4.037 marks? The figure has changed because the correlation between books and attend was not equal to zero. Because students who read more books also attend more lectures, when attendance is controlled for we find that reading more books helps, but not as much as we previously calculated. The unstandardised coefficient tells us how much reading each extra book would help, if everyone attended the same number of lectures.

We get a similar story when we look at the standardised coefficients. Now we are not looking just at the correlation between one independent variable and the dependent variable; instead we are looking at what the correlation would be if the other independent variables were held constant. This shows that the correlation between books and grade has dropped – from 0.492 in the previous chapter when books was the only independent variable to 0.346 when attend is included.

2.8 Variable entry

2.8.1 Hierarchical variable entry

So far, when we have been assessing the contribution of an independent variable we have assessed it at the same time as all of the other variables. However, sometimes we may be much surer about the causal importance, or hierarchy, of our variables. In this section we will show how a hierarchy of variables can be entered into a multiple regression analysis. (We will expand on notions of causality in Chapter 5.)

Table 2.7 shows a dataset that has four variables: sex, age, extroversion and car; car is a variable that refers to the average number of minutes per week a person spends looking after their car.

TABLE 2.7 Dataset 2.2

sex	age	extroversion	car
1	55	40	46
1	43	45	79
0	57	52	33
1	26	62	63
0	22	31	20
0	32	28	18
0	26	2	11
1	29	83	97
1	40	55	63
0	30	32	46
0	34	47	21
1	44	45	71
1	49	60	59
1	22	13	44
0	34	7	30
1	47	85	80
0	48	38	45
0	48	61	26
1	22	26	33
0	24	3	7
0	50	29	50
0	49	60	54
1	49	47	73
0	48	18	19
0	29	16	36
0	58	36	31
1	24	24	71
0	21	12	15
1	29	32	40
1	45	46	61
1	28	26	45
0	37	40	42
1	44	46	57
0	22	44	34
0	38	3	26
0	24	25	47
1	34	43	42
1	26	41	44
1	26	42	59
1	25	36	27

A person may project their self-image through themselves or through objects that they own, such as their cars. Therefore a theory could be developed which predicted that people who score higher on a measure of extroversion are likely to spend more time looking after their cars. An advocate of such a theory might first carry out a correlation analysis between the extroversion and car variables. Suppose they found that the correlation between extroversion and car was 0.671 (equivalent to $R^2 = 0.45$; that is, 45% of the variance in car has been explained by extroversion).

In our role as critical reviewers of this research, we could argue two alternative explanations. First, from the previous literature we know that males tend to score higher on extroversion scales than females *and* we can also make the assumption that males are likely to spend more time looking after their car than females. Second, we know that older people tend to have lower scores on extroversion scales and also spend less time looking after their car. Basically, it could be argued that the correlation between extroversion and car ($r = 0.671$) does not reflect the true relationship between these two variables because extroversion also contains information about sex and age. What we would want is to estimate the effect of extroversion on car while removing, or controlling for, the effects of sex and age.

To do this the researcher could carry out a regression analysis using the sex (see Chapter 3 for details of how categorical variables may be coded as independent variables), age and extroversion variables as independent variables, and car as a dependent variable. The results of this analysis are shown in Table 2.8. The researcher could then argue that extroversion has a beta value of 0.441, and that therefore extroversion accounts for ($0.441^2 = 0.19$) 19% (rather than 45%) of the variance in car.

TABLE 2.8 *Results of regression analysis with* car *as the dependent variable and* sex, age *and* extroversion *as independent variables*

	Slope (*b*)	Std error of slope	Standardised slope (beta)	*t*	Sig.
Constant	11.426	7.308		1.563	0.127
sex	20.038	4.650	0.488	4.309	0.000
age	0.154	0.206	0.085	0.749	0.459
extro.	0.463	0.130	0.441	3.574	0.001

We could still argue that this was not the most appropriate answer as the contributions of sex, age and extroversion have been assessed simultaneously. We will argue that age and sex must be considered to be primary causes of car, but what we are really interested in is the effect that extroversion has *above and beyond* the effects of age and sex — that is to say, we want to assess the effects of extroversion after the effects of age and sex have been removed. (The idea of causal order will be discussed further in Chapter 5).

To do this we carry out two separate regression analyses, first with only sex and age as independent variables, and second with sex, age and extroversion as independent variables. We can then compare the values of R^2, and carry out a significance test to see if the increase is statistically significant, or is likely to have arisen by chance.

With most of the software available to you, this analysis can be done automatically. However, in some programs the process is a little tricky and you may prefer to use the following formula to calculate a value for *F* (the

ANOVA test statistic). The value for F can be compared against the critical values found in Appendix 3:

$$F = \frac{(R_2^2 - R_1^2)/(k_2 - k_1)}{(1 - R_2^2)/(N - k_2 - 1)}$$

where R_1^2 is the value of R^2 for the first (smaller) model, R_2^2 is the value of R^2 for the second (larger) model, k_1 is the number of independent variables in the first model and k_2 is the number of independent variables in the second model. df for the F-value is equal to $k_2 - k_1$, $N - k_2 - k_1$. If we now carry out this analysis, we find the results for R and R^2, which are shown in Table 2.9.

TABLE 2.9

Model	R	R square
sex age	0.714	0.510
sex age extro.	0.799	0.638

If we substitute the values in Table 2.9 into the equation shown above to calculate F, we find that:

$$F = \frac{(R_2^2 - R_1^2)/(k_2 - k_1)}{(1 - R_2^2)/(N - k_2 - 1)}$$

$$= \frac{(0.638 - 0.510)/(3 - 2)}{(1 - 0.638)/(40 - 3 - 1)}$$

$$= \frac{0.128/1}{0.362/36}$$

$$= \frac{0.128}{0.01005}$$

$$= 12.7$$

We also find that the numerator df is equal to $k_2 - k_1 = 2 - 1 = 1$, and the denominator is equal to $N - k_2 - k_1 = 40 - 3 - 1 = 36$.

Therefore $F = 12.7$, $df = 1, 36$, and if we consult the tables in Appendix 3, we find that $p < 0.05$. This means that the increase is significant, and extroversion does predict car, above and beyond sex and age. However, if we look at the actual increase in R^2, we find that it increases by 0.128, approximately equal to 13%, and therefore extroversion is accounting for less of the variance than we had at first thought.

2.9 Methods of variable entry

There are techniques available in most statistical packages that allow regression models to be built in a series of steps, adding or removing one independent variable at a time. The main difference between these models and the hierarchical models discussed above is that computer packages use statistical criteria to determine the usefulness of a particular variable, rather than psychological theory. However, the general aim is the same – to find a *parsimonious* model. A parsimonious model is one that explains the most variance in the dependent variable containing the fewest number of independent variables.

The *backward* technique starts off with a model that includes all of the independent variables. The parameters are estimated, and any variables that do not have significant parameters, at a pre-specified level (usually 0.10), are removed from the equation, and it is then re-estimated. Any non-significant variables are removed, and the equation is estimated (again). This process continues until no more independent variables are significant.

The *forward* technique is very similar, except that it starts with none of the variables in the equation. The variable with the highest value of standardised beta, which is also significant (usually at $p < 0.05$), will be added into the equation. The variable that would have the next highest value of standardised beta is then assessed, to see if it would have a significant value. This continues until no more variables are significant.

Stepwise regression is a combination of these two, adding variables when they are significant, and removing them when they are not significant.

This family of techniques (forward, backward and stepwise regression) has a large number of problems associated with it, and should be used with extreme caution. The basic problem with stepwise regression techniques is that we are asking the computer to make a decision regarding which variables are important, when the computer has no idea about the theory that may determine which variables are important.

An analogy is that if stepwise regression were used to pack your suitcase, it would select the item of clothing that seemed to be the best – a pair of trousers, for example. Then it would examine which items of clothing fitted, *based on what clothes were already packed*. Underwear does not fit well when trousers are in first, so stepwise regression would reject underwear, as it does not fit the model.

There was considerable discussion on the newsgroup sci.stat.consult regarding the problems with stepwise regression, and the comments were collated into a FAQ[1] on the problems with stepwise regression (Ulrich, 1997).

One of the main problems with stepwise regression is that it yields R^2-values that are badly biased high. Stepwise regression trawls through a potentially large number of independent variables and selects those that are significant predictors. A proportion of variables equal to alpha (usually 0.05) will be significant in random data, and each of these will increase the value of R^2. The adjusted R^2 does not compensate for this, as it only adjusts for the number of variables that are in the model; it does not take account of how many were potentially in the model and were rejected.

Additionally, the significance values of both the R^2 and the individual beta values are incorrect. Because the calculation of the p-values depends on the number of variables that are being evaluated, removing a variable from the analysis will alter the p-values.

One final problem identified in the FAQ with the use of stepwise selection procedures is that, because of the large number of models that are estimated, they use an awful lot of paper when you come to print them out.

Cohen and Cohen (1983) criticised stepwise regression procedures for their lack of theoretical input:

> We take a dim view of the ... use of stepwise regression in explanatory research for various reasons ... but mostly because we feel that more orderly advance in the behavioral sciences is likely when researchers, armed with theories, provide a priori theoretical ordering that reflects causal hypotheses rather than when computers order IVs post and ad hoc for a given sample. (p. 124)

They are further unimpressed by stepwise regression because the same researchers collecting the same data from the same population are not particularly likely to get the same answer:

> Probably the most serious problem with the use of stepwise regression is when a relatively large number of IVs is used. ... A related problem with the free use of stepwise regression is that in many research problems the ad hoc order produced from a set of IVs is likely not to be found in other samples from the same population. (p. 124)

Note

1 FAQ (pronounced 'fack') stands for frequently asked questions. It is a set of questions that are commonly asked on Internet newsgroups and mailing lists, and the answers that have been given in the past.

Further Reading

See Cohen and Cohen (1983) *Applied multiple regression/correlation analysis for the behavioural sciences*, Chapter 3, and Pedhazur (1997) *Multiple regression in behavioral research: explanation and prediction*, Chapter 7.

3 Categorical independent variables

3.1 Introduction

There is a perceived split in psychological research between 'experimental' psychologists, who use techniques such as ANOVA to examine mean differences (generally across experimental conditions), and 'correlational' psychologists who analyse their data using regression- and correlation-based statistical techniques. What many psychologists do not realise is that both these approaches to research are doing the same thing, but just going about it in a different way – the difference is more perceived than real.

How did it come about that these two separate approaches developed? Before the days of modern computers, doing statistical analysis was time consuming and technically demanding. To calculate a multiple regression analysis would take somewhere between a couple of hours and several days, and many other statistical techniques were virtually unusable,[1] simply because they would take too long.

Because the calculations took so long, there was considerable pressure to find an easier way either to do the calculations by simplifying them, or to achieve the same results. Two examples of this that you may come across are the phi correlation, and the point–biserial correlation. These are both simplifications of the correlation formula, which can be used when either both variables (phi) or one variable (point–biserial) can only take on two values.

3.1.1 Categorical data: a special case

Up to now we have (implicitly) considered only independent variables which are continuous, that is where there is a scale and individuals can score somewhere along that scale, from low to high. Often in psychological research we have variables which are categorical, or nominal. A categorical variable comprises a number of categories, rather than a continuous scale. The most common example in psychological research occurs when a traditional experiment is carried out. In this we look for differences between two or more groups.

If the independent variables are all categorical then there is a simple way to deal with the equations for regression analysis. The simplified method involves partitioning the sums of squared deviations from the mean. This approach was mentioned earlier, namely analysis of variance (ANOVA). The ANOVA approach became the standard way to analyse data when the independent variable was categorical, because it reduced the complexity of the calculations. When computer programs designed for statistical analysis were introduced, they too separated the ways of carrying out ANOVA and regression; even though having a simple method was now irrelevant, because computers did the work, people did not know, or remember, that they were doing regression. The existence of the two methods led to a belief amongst many scientists that regression and ANOVA were different things; the simple fact is that ANOVA is a special case of regression analysis, one where the independent variables are categorical rather than continuous. Recently, ways of dealing with this type of calculation have started to change: for example, SPSS has developed a general linear model (GLM) procedure which combines regression and ANOVA.

In this chapter we will show how to analyse data that contain categorical independent variables using regression analysis. Although ANOVA represents an easier method of carrying out the analysis, we believe that using regression does have some advantages, under some circumstances. (We are not saying that we would always use regression – it is often quicker and easier to carry out ANOVA, but it can be useful to know about regression approaches to ANOVA.) One advantage is that regression is conceptually clearer: there is no distinction made between categorical variables and continuous independent variables. There are good reasons for including continuous variables in many analyses that use ANOVA (Taylor & Innocenti, 1993) but because researchers consider themselves to be 'ANOVA researchers' they do not use the continuous variables. A second, and more important, advantage is that regression avoids the necessity of having to categorise continuous variables to make the data 'fit' the ANOVA model. When researchers are interested in a continuous independent variable, they are often tempted to split it into 'high', 'medium' and 'low' so that ANOVA can be used. By making this split they are throwing away information contained within their hard-earned data, and may even be opening themselves up to spurious interactions (Maxwell & Delaney, 1993). A third advantage is that regression forces the analyst to consider how the independent variables may interact in terms of their influence on the dependent variable. Computer programs using ANOVA will automatically estimate all possible interactions of variables – which leads to what has been called 'data fishing': just try everything and see what you can find that is significant. Using regression analysis forces the researcher to specify and calculate specific interactions in advance, and this process in turn leads to a more theory-driven procedure. By avoiding 'data fishing' we avoid inflating our type I (false/positive) error rate above the nominal value of 0.05. Finally an understanding of regression analysis makes it easier to understand what is happening, and why, when GLM procedures are used.

This chapter looks at how regression can be used to analyse data from a simple between-groups experiment with two conditions – normally an independent-samples t-test would be used to analyse these sorts of data. Then we show how to use regression as an alternative to one-way ANOVA and analysis of covariance (ANCOVA). (In Chapter 5 we show how to use regression analysis to analyse data from a 2 × 2 experimental design.)

3.2 The t-test as regression

The first example of using regression analysis to analyse a categorical independent variable is an experiment that is designed to test the effects of mnemonics in memory. The experiment has two groups: a control group, who receive no instruction, and an experimental group, who receive instruction in the use of mnemonics as a memory aid. The experiment has one dependent variable, the number of items correctly remembered. We want to compare the mean number of items remembered by each of the two groups to see if there is any difference between the two groups. Usually, these data would be analysed using a *t*-test, but we will show in this chapter how the same results can be obtained using regression analysis.

Now let us think about the analysis in terms of a regression analysis: we have one independent variable, which is group membership (control group, coded as 0, and experimental group, coded as 1), and one dependent variable, which is the number of correct responses. As you should remember, the intercept in a regression equation is the value of the dependent variable when the independent variable is equal to zero. In this case the independent variable is equal to zero in the control group and therefore the intercept will be equal to the mean for that group. The slope in a regression equation is the increase in the score of the dependent variable when the independent variable increases by one unit. Obviously, if we add one to zero, the result is one, which is the coding value of the experimental group. The beta, or slope, coefficient will be equal to the difference between the two means. To try and make this point clearer, let us think about some possible results of this experiment and how they would look when plotted on a line graph. Table 3.1 shows the results of three such experiments.

TABLE 3.1

	Control group mean (0)	Experimental group mean (1)
Experiment 1	10	10
Experiment 2	10	20
Experiment 3	10	50

The results of the three experiments are shown graphically in Figures 3.1, 3.2 and 3.3. In each case we have assigned the value 0 to the control group and 1 to the experimental group. In Experiment 1 shown in Figure 3.1, the means for the control and experimental groups are the same (10.00) and hence the difference between them is zero. This zero difference is reflected in the slope of the line between the two means. The line is flat, which means that the slope is zero. In Figure 3.2 there is a small difference between the group means, and this small difference results in a shallow slope. The steepest slope is in Figure 3.3 where there is a substantial difference between the means.

We can also calculate the slope for each of the three experiments, demonstrated above. This is done by finding how much the dependent variable (in

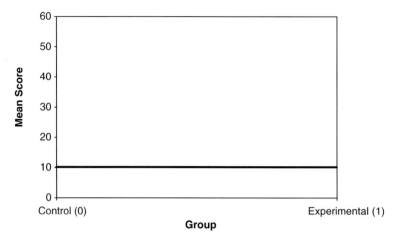
FIGURE 3.1

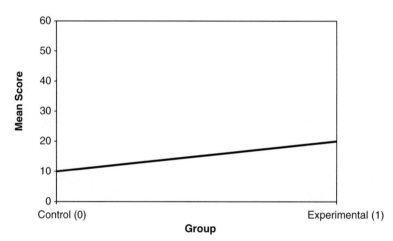
FIGURE 3.2

this case the number of items remembered) changes, when the independent variable increases by one unit. In this example the independent variable is the group, and the difference between the control group and the experimental group is one unit. This means that *the slope is equal to the difference between the means*. In Experiment 1, the difference between the means is zero, and the slope was also equal to zero. In Experiment 2, the difference between the means was 10, and the slope was equal to 10. In Experiment 3, the difference between the means was 40, and the slope was therefore equal to 40.

The above example illustrates the principle of a slope being equivalent to a difference in means between two groups. In the following example we will see how this translates to a whole set of data, rather than simply two means.

Table 3.2 shows the data from an experiment designed to assess the use of mnemonics in memory. Two groups of 10 participants took part in the experiment. The participants in the control group (labelled 0 in the table) were

44 APPLYING REGRESSION AND CORRELATION

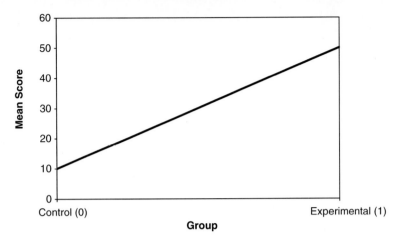

FIGURE 3.3

TABLE 3.2 *Data from experiment on memory (dataset 3.1)*

group	score
0	10
0	8
0	13
0	9
0	10
0	13
0	10
0	9
0	11
0	8
1	13
1	15
1	10
1	12
1	11
1	12
1	11
1	15
1	16
1	11

asked to try to remember as many household objects from a list with which they were presented. The participants in the experimental group (labelled 1) completed the same task but had been previously shown how to use a mnemonic to improve the performance of their memory. All participants were then asked to try to remember as many items from the list as they could. The number of items they remembered is labelled score.

The standard way of analysing these data would be to carry out a *t*-test. Table 3.3 shows the summary information, such as the mean and standard deviation, for each of the two groups.

TABLE 3.3

group	N	Mean	Std deviation	Std error mean
0	10	10.10	1.79	0.57
1	10	12.60	2.07	0.65

The usual statistic that is reported is t, along with its associated significance. In this example $t = 2.89$, $df = 18$, and $p = 0.01$. This result shows that there is a significant difference between the mean number of objects recalled by the two groups. We can see from the summary statistic that the experimental group is scoring higher than the control group.

Table 3.4 shows the output from the regression analysis for the data collected from the memory experiment. The intercept is the same as the mean value for the control group shown in Table 3.3. The slope is the same as the difference between the means. Finally the t-value and significance are equal to those found by the t-test.

TABLE 3.4 *Output from regression analysis*

	Slope (b)	Std error of slope	Standardised slope (beta)	t	Sig.
Constant	10.100	0.611		16.518	0.000
group	2.500	0.865	0.563	2.891	0.010

3.3 ANOVA as regression

In this section we will examine how regression analysis can be used instead of a one-way ANOVA. It is possible to use regression with repeated measures and mixed designs, but the techniques do become a little complex and go beyond the scope of this text. If you are interested in this technique, we recommend you see Pedhazur (1982)[2] or Rutherford (2000).

3.3.1 Coding schemes for categorical data

There is a specific kind of problem that occurs when we have three or more groups. If we were to extend the logic of our previous example, we would have an independent variable, which would have three values, 0, 1 and 2. Unfortunately, if we were to enter this variable as an independent variable into our regression equation we would find that the analysis treated the categorical variable as if it were continuous, and therefore our result would be wrong. Instead we need to recode the independent variable into a series of variables which can then be used in the regression analysis. There are several ways of coding categorical data and here we will consider two of the most common, which we call dummy coding and effect coding. (Watch out, because some textbooks use slightly different terms for these. Hutcheson and Sofreniou (1999), for example, use the term dummy coding to refer to any coding scheme

46 APPLYING REGRESSION AND CORRELATION

for categorical data, the term indicator coding for what we call dummy coding, and the term deviation coding for what we call effect coding. SPSS (SPSS, Inc., 1999), in its GLM procedure, uses the terms simple coding and deviation coding.) But before we go into this, we will describe a hypothetical experiment and analyse the results using the traditional ANOVA approach.

3.3.1.1 ANOVA approach

As an example we will use an extension of the experiment on memory we looked at in the previous example. We now have three groups, a control group (coded 0), a mnemonic group (coded 1) and a second experimental group (coded 2) (Table 3.5). The second experimental group had the room infused with the fragrance of rosemary, an aromatherapy oil, with reputed memory enhancing qualities.

TABLE 3.5 Dataset 3.2a

group	score
0	10
0	8
0	13
0	9
0	10
0	13
0	10
0	9
0	11
0	8
1	13
1	15
1	10
1	12
1	11
1	12
1	11
1	15
1	16
1	11
2	8
2	10
2	9
2	9
2	6
2	6
2	9
2	13
2	10
2	9

The summary statistics are shown in Table 3.6 and the results of the ANOVA approach are shown in Table 3.7. Table 3.8 shows the results of *post hoc* tests using the least significant difference (LSD) test.

TABLE 3.6 *Summary statistics for three-group memory experiment using ANOVA*

group	N	Mean	SD	Std error	95% CI for mean	
					Lower	Upper
0 (Control)	10	10.10	1.79	0.57	8.82	11.38
1 (Mnemonics)	10	12.60	2.07	0.65	11.12	14.08
2 (Aromatherapy)	10	8.90	2.02	0.64	7.45	10.35
Total	30	10.53	2.46	0.45	9.61	11.45

TABLE 3.7 *Results for three-group memory experiment using ANOVA*

	Sum of squares	df	Mean square	F	Sig.
Between groups	71.267	2	35.633	9.233	0.001
Within groups	104.200	27	3.859		
Total	175.467	29			

TABLE 3.8 Post hoc *tests for three-group memory experiment using LSD*

(I) group	(J) group	Mean difference (I−J)	Sig.
0 (Control)	1 (Mnemonics)	−2.50	0.008
	2 (Aromatherapy)	1.20	0.183
1 (Mnemonics)	0 (Control)	2.50	0.008
	2 (Aromatherapy)	3.70	0.000
2 (Aromatherapy)	0 (Control)	−1.20	0.183
	1 (Mnemonics)	−3.70	0.000

From the information in these three tables we can make the following observations. First, the summary statistics indicate that the mnemonics condition (1) has the highest mean recall of the three groups, and that the control (0) and aromatherapy (2) are relatively similar (Table 3.6). The results of the ANOVA approach (Table 3.7) confirm that the means are not equal. The *post hoc* tests (Table 3.8) show that there are significant differences between the mnemonics and control conditions, and between the mnemonics and aromatherapy conditions. The mnemonics condition has the highest mean, and the difference between the means for the control and aromatherapy conditions is not significant.

3.3.1.2 *Dummy variable coding*

We can analyse the same data using regression after *dummy coding* the independent variable. In dummy coding one group is considered to be the reference group (the control condition in this case), and new dummy variables are created to identify which condition the other subjects are in. For the memory experiment two new variables need to be created. These two new variables refer to each group in the original independent variable *except* for the reference group. So the first new variable, group_1, should have only zero's to indicate those subjects not in the mnemonics condition (Group 1) or ones to indicate those subjects present in the mnemonics condition (Group 1). The same principle is used to code the second variable, group_2. Again this variable contains only zeros or ones: the ones indicate membership of the aromatherapy condition (Group 2) and the zeros membership of any other condition.

The dataset shown in Table 3.5 is reproduced in Table 3.9 with two new variables – the dummy-coded variables group_1 and group_2.

TABLE 3.9 *Dataset from three-group memory experiment with dummy-coded variables (dataset 3.2b)*

group	score	group_1	group_2
0	10	0	0
0	8	0	0
0	13	0	0
0	9	0	0
0	10	0	0
0	13	0	0
0	10	0	0
0	9	0	0
0	11	0	0
0	8	0	0
1	13	1	0
1	15	1	0
1	10	1	0
1	12	1	0
1	11	1	0
1	12	1	0
1	11	1	0
1	15	1	0
1	16	1	0
1	11	1	0
2	8	0	1
2	10	0	1
2	9	0	1
2	9	0	1
2	6	0	1
2	6	0	1
2	9	0	1
2	13	0	1
2	10	0	1
2	9	0	1

The data are now ready to analyse using regression. If group_1 and group_2 are entered as independent variables and score as the dependent variable, a regression analysis will produce a typical ANOVA table and regression coefficients with associated test statistics. These are shown in Table 3.10 and Table 3.11 respectively.

TABLE 3.10 *ANOVA table from regression analysis of three-group memory experiment using dummy variable coding*

	Sum of squares	df	Mean square	F	Sig.
Regression	71.267	2	35.633	9.233	0.001
Residual	104.200	27	3.859		
Total	175.467	29			

TABLE 3.11 *Parameter estimates from regression analysis of three-group memory experiment using dummy variable coding*

	Slope (*b*)	Std error of slope	Standardised slope (beta)	*t*	Sig.
Constant	10.100	0.621		16.258	< 0.001
group_1	2.500	0.879	0.487	2.846	0.008
group_2	−1.200	0.879	−0.234	−1.366	0.183

The ANOVA table shows that the result is significant, meaning that the null hypothesis (of no mean differences) can be rejected. As with the ANOVA result in Table 3.7 we cannot tell where the differences lie, only that there is at least one pair of means that are significantly different. Note that the values for *F* and the significance of *F* in the regression analysis (Table 3.10) and the ANOVA table (Table 3.7) are also identical.

To understand the nature of the differences we need to examine the regression coefficients, or slopes, associated with each of the independent variables. This process is much the same as interpreting the *post hoc* tests in the traditional ANOVA approach outlined earlier. As with the *t*-test, these slopes indicate the difference between the mean value for each group and the reference group (in this case the control group). The regression coefficient for group_1 is statistically significant but the regression coefficient for group_2 (Table 3.11) is not statistically significant. Therefore the means of the control and mnemonics condition (group_1) are significantly different, but the means of the control and aromatherapy condition (group_2) are not significantly different. The sign of the slope gives us information about the nature of the mean difference. If the slope is positive, as with group_1, the mean of that condition is higher than that of the reference group. Correspondingly, a negative slope, as with group_2, implies that the mean of that condition is lower than the mean of the reference group (not significantly lower, in this case). This interpretation of the results is the same interpretation that was reached when the data were analysed using ANOVA and *post hoc* tests.

At this point we should admit that we could have been more punctilious regarding the analysis. We have to admit that we have been capitalising on chance. That is to say, we have been increasing the chances of a type I error (falsely rejecting the null hypothesis). The reason for this increase is that these tests are not corrected for multiple comparisons. Every time we test a null hypothesis, if that null hypothesis is correct, we have a 0.05 chance (1 in 20) of finding a significant result, and therefore making a type I error. If we do lots of comparisons, we are likely to find some significant differences just by chance – it is as if we are given three rolls of the die in a game of snakes and ladders, and then allowed to pick which of the three we wanted. If we do three comparisons (control against Group 1 – mnemonic; control against Group 2 – rosemary and Group 1 – mnemonic, against Group 2 – rosemary). The probability of a type I error is increased from 0.05, by almost three times, to almost 0.15 – about 1 in 6. We need to carry out some sort of correction to get our type I error rate back down to 0.05.

The Bonferroni correction is the simplest form of correction to apply to these significance tests. We will cover it very briefly here; more applications are shown in Roberts and Russo (1999) or Maxwell and Delaney (1990). The Bonferroni correction, named after the Italian statistician Carlo Bonferroni,[3] is done by dividing the original significance level, in this case 0.05, by the number of comparisons that are being made. In this analysis we make two comparisons (control and mnemonics, control and aromatherapy). So to retain an overall type I error rate of 0.05, we need to test the statistical significance of each slope at the 0.025 level (0.05/2), rather than 0.05. This does not change the interpretation; the probability value for group_1 (0.008) remains less than 0.05 and the probability value for group_2 (0.183) remains greater than 0.05.[4]

Note that we did not make a third comparison between mnemonic and rosemary. To do this we would need to rerun the analysis using one of the other variables as a reference variable, and we would also have to correct for three comparisons, rather than two.

3.3.1.3 Effect coding

In the previous example it was clear that there was a particular reference group that could be used to make comparisons. However, we do not always have such a reference group. Table 3.12 shows the levels of stress in five different types of educator; primary school teacher (1), secondary school teacher (2), college lecturer (3), old university lecturer (4) and new university lecturer (5).[5]

To analyse the differences between the mean stress experienced by each of these groups we could, as before, use dummy coding. However, there is no obvious reference group. Instead we could use indicator coding to compare the scores of a group with the mean score for all groups.

Effect coding is very similar to dummy coding. The main difference between these two types of coding is that with dummy coding one group is used as a reference group and is given a zero for every score, but with effect coding one group is arbitrarily chosen and given the value -1 for every score. When we do this the number of variables is still one less than the number of groups. Table 3.13 shows how the group variable would be recoded into four variables.

The results of the regression analysis are shown in Table 3.14 and Table 3.15. The ANOVA table (Table 3.14) shows that the overall result is significant, and that the null hypothesis of equivalent means can be rejected. The

TABLE 3.12 *Dataset 3.3*

job	stress
1	71
1	67
1	67
1	67
1	79
1	46
1	76
1	82
1	55
1	64
2	30
2	44
2	58
2	67
2	92
2	74
2	56
2	58
2	51
2	46
3	33
3	64
3	54
3	70
3	56
3	97
3	66
3	77
3	76
3	53
4	24
4	21
4	57
4	52
4	52
4	21
4	66
4	43
4	32
4	79
5	35
5	50
5	41
5	49
5	71
5	34
5	70
5	59
5	46
5	68

TABLE 3.13 *Effect coding for five-group stress study (incorporated into dataset 3.3)*

group	group_1	group_2	group_3	group_4
1	1	0	0	0
2	0	1	0	0
3	0	0	1	0
4	0	0	0	1
5	−1	−1	−1	−1

TABLE 3.14 *ANOVA table for stress study*

	Sum of squares	df	Mean square	F	Sig.
Regression	3 391.48	4	847.87	3.23	0.020
Residual	11 787.40	45	261.94		
Total	15 178.88	49			

TABLE 3.15 *Parameter estimates for effect-coded stress analysis*

	Slope (b)	Std error of slope	Standardised slope (beta)	t	Sig.
Constant	57.320	2.289		25.043	0.000
group_1	10.080	4.578	0.366	2.202	0.033
group_2	0.280	4.578	0.010	0.061	0.951
group_3	7.280	4.578	0.264	1.590	0.119
group_4	−12.620	4.578	−0.458	−2.757	0.008

regression coefficients (Table 3.15) show where these differences arise. The constant shows the overall mean for all of the groups (57.3) and the slope coefficients show the amount that each group differs from the overall mean. Only group_1 (primary school teachers) and group_4 (old university lecturers) have scores that are significantly different from the mean score, primary school teachers having significantly more stress and old university lecturers having significantly less.

This analysis excluded new university lecturers because they were used as the reference group. To compare the stress levels of this group to the mean we would recode the data to use a different reference group.

3.3.1.4 Dummy coding and the analysis of change

We have, we hope, shown that ANOVA is simply a special case of regression analysis, and that any analysis that can be done with a one-way ANOVA approach can also be carried out using regression. We will demonstrate in a later chapter how it is possible to code for factorial ANOVA including interaction effects. At this point the cynical reader may ask what we have achieved — why not continue to use ANOVA? The answer is because there are some circumstances where there are clear advantages to using regression rather than ANOVA. (The approach that we describe in the following section is the equivalent of ANCOVA.)

A clear example of these advantages is shown in the analysis of change scores, for instance in experiments examining context-dependent implicit memory. Such experiments usually involve a two-stage process. In stage 1 the participants are randomly allocated to a particular experimental condition (warm or cold, smell present or absent, etc.), the independent variable. Then they are asked to examine words, and rate the words on some criterion, for example 'emotionality' shown as an example in Table 3.16. Participants are not told, or led to believe, that they are to learn these words. This process of rating words is called priming, and words that have been included in the rating process are called *primed* words. They have been learned, implicitly, without the participant being aware that they have been learned.

TABLE 3.16 *Example words to rate*

Word	Emotionality of word
Elephant	Very : fairly : slightly : not at all
Carrot	Very : fairly : slightly : not at all
Tablecloth	Very : fairly : slightly : not at all
Pugnacious	Very : fairly : slightly : not at all
Automatic	Very : fairly : slightly : not at all

After a break, participants are asked to take part in what they believe to be an unrelated experiment on word completion. They are given letters and spaces, and asked to fill in the blank spaces to complete words. These blanks, with answers, are shown in Table 3.17. While the participants are filling in the blanks, the context in which they carried out the rating process is recreated (e.g. the temperature, or the smell). There are two types of words to complete: some were included in the list of words that were rated (primed words, labelled as P in the table) and some were not (unprimed words, labelled as U in the table). The dependent variable is the number of *primed* words; primed words are those words that appeared in the previous part of the study.

TABLE 3.17 *Word completion task from context-dependent implicit memory study*

Letters	Word	Type
_l___h__t	Elephant	P
__t__f_ll	Waterfall	U
c__o_	Carrot	P
_r__s__	Trousers	U
__bl___o_h	Tablecloth	P
__e__o_e	Telephone	U
_g__c__u_	Pugnacious	P
_iv__s__y	University	U
_t_m___c	Automatic	P
__bb__e	Cabbage	U

At this point you may be thinking that filling in the blanks to complete words is a skilled task and that people with a larger vocabulary, or better word recognition skills, would perform better. If this were the case we would want to control statistically for the effect of vocabulary as we anticipate it has a significant influence on the dependent variable. Fortunately we have a measure of vocabulary – the number of *unprimed* words that were completed, that is, the number of words completed that were not in the first part of the study. If in some way we take the number of unprimed words into account, we should be able to get a more sensitive measure of the effect of our independent variable. Table 3.18 shows the data from a hypothetical experiment in which these data have been collected. The variable group contains two experimental groups, a control group (0) and an experimental group (1), unprim is the number of unprimed words the participant successfully completed, and prim is the number of primed words the person successfully completed. The table also includes diff, the difference between prim and unprim.

There are three possible strategies that can be employed to analyse this data.

The first approach is to calculate the difference between the two scores. By subtracting the number of unprimed words from the number of primed words, we will get a score that reflects the 'improvement' that occurred for each group as a result of our experimental manipulation. We have included these data in the table, under the variable name diff. The analysis could then be carried out with a t-test (or, of course, regression) to compare the two groups. When we do this analysis, we find that $t = 1.95$, $df = 48$ and $p = 0.056$. We would therefore fail to reject the null hypothesis, and could not conclude that our experimental manipulation was having an effect.

A second approach would be to use a mixed ANOVA, where the prim vs unprim words are treated as a repeated-measures factor, and group is treated as an independent-groups factor. In this case we would hypothesise that there would be an interaction effect between the independent variables – the effect of the manipulation should have an effect in the experimental group, but not in the control group. The graph shown in Figure 3.4 seems to show that an interaction effect is occurring: in the experimental group there is a large difference between the primed and unprimed words, whereas in the control group the difference is much smaller. The results shown in Table 3.19 demonstrate that this interaction is (only just) non-significant (at the 0.05 level), and this interpretation is the same interpretation that was reached using the t-test. (In fact this was exactly the same test, the p-values are the same, and the value of F is equal to the square of the t-value.)

The problem with both of these methods of testing the effect is that they make an assumption that the correlation between the primed score and the unprimed score is 1, or at least very close to 1. If the correlation is lower, then the results can be misleading. A third and much better approach is to use regression analysis, entering the prim scores as a dependent variable, and the unprim scores as an independent variable along with the group variable. The vocabulary level is therefore controlled for, taking into account the correlation between primed and unprimed scores.

The results of this analysis using regression are shown in Table 3.20. This shows that the effect of the experimental manipulation is now significant at the 0.05 level.

TABLE 3.18 *Dataset 3.4*

group	unprim	prim	diff
0	87	70	−17
0	87	87	0
0	80	86	6
0	82	96	14
0	70	52	−18
0	87	123	36
0	75	59	−16
0	98	90	−8
0	83	86	3
0	95	77	−18
0	71	55	−16
0	76	95	19
0	62	84	22
0	68	59	−9
0	86	95	9
0	76	58	−18
0	85	70	−15
0	83	122	39
0	60	129	69
0	73	108	35
0	66	52	−14
0	69	33	−36
0	87	89	2
0	68	93	25
0	80	79	−1
1	68	92	24
1	81	125	44
1	80	71	−9
1	102	80	−22
1	73	121	48
1	93	120	27
1	78	117	39
1	67	102	35
1	65	79	14
1	91	101	10
1	73	71	−2
1	78	42	−36
1	51	73	22
1	67	132	65
1	73	99	26
1	79	126	47
1	91	121	30
1	85	94	9
1	88	107	19
1	83	77	−6
1	84	96	12
1	75	65	−10
1	83	56	−27
1	96	167	71
1	74	92	18

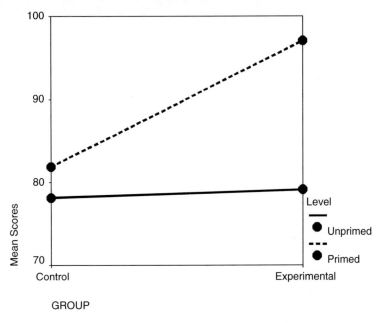

FIGURE 3.4 *Graph showing interaction effect in mixed ANOVA*

TABLE 3.19 *Results of mixed ANOVA to analyse memory experiment*

Source	F (df = 1, 48)	Sig.
group	3.52	0.067
factor1	8.9	0.005
factor1 × group	3.8	0.056

TABLE 3.20 *Parameter estimates for regression analysis, using* unprim *and* group *as independent variables*

	Slope (*b*)	Std error of slope	Standardised slope (beta)	*t*	Sig.
Constant	35.792	27.717		1.291	0.203
unprim	0.590	0.349	0.230	1.692	0.097
group	14.594	7.239	0.274	2.016	0.049

The alert reader may have noticed that the difference in the significance values between the three methods of analysis was very small – the first two techniques only just failed to achieve significance, and the regression technique only just achieved significance. The regression approach is therefore a slightly more powerful test of the hypothesis. The power difference in this example was small, but this is due in part to the small sample size – the larger the sample size that is used, the greater the difference in power between the two techniques. Additionally, the smaller the correlation between the primed and unprimed words, the larger the power difference will be. If the correlation is sufficiently low, using either the difference score method or the mixed ANOVA method can lead to a lower probability of finding a significant result than would be achieved by simply ignoring the unprimed score variable.

We hope that this section has further illustrated the flexibility of using regression. It can be used to analyse either non-experimental or experimental data, it can deal with both continuous and categorical data, and it can be used to indicate the direction and magnitude of particular effects.

Notes

1 One example of this which we will encounter in Chapters 6 and 8 is maximum likelihood estimation. This was developed in the 1930s, but was unusable until the 1960s when computers became fast enough.

2 Note that this is the 2nd edition of the book; the 3rd edition does not contain the technique.

3 Although, curiously, he did not actually develop the technique. Even more curiously, this seems to be a common occurrence in statistics. Stigler (1980) has referred to this as 'Stigler's law of eponymy', although in order to satisfy the law, he should not have proposed it. And he didn't.

4 An alternative but equivalent procedure is to multiply each of the probability values found by the number of comparisons.

5 By this we do not mean the age of the university lecturer, rather we mean the type of university. In the UK, there is a distinction between new universities and old universities. New universities were usually previously called polytechnics, but attained university status in 1992. Old universities were universities prior to this date, the youngest being founded in the early 1970s. We should perhaps point out that one of us is currently employed by a new university, and the other by an old one.

Further reading

Cohen and Cohen (1983) *Applied multiple regression/correlation analysis for the behavioral sciences*, Chapter 5, and Pedhazur (1997) *Multiple regression in behavioral research: explanation and prediction*, Chapter 11, both provide detailed descriptions of methods of dealing with categorical data, including some that we do not include here.

PART II

I NEED TO DO REGRESSION ANALYSIS NEXT WEEK

4 Assumptions in regression analysis

4.1 Introduction

Up to this point in the book we have simply taken our data and 'thrown' a data analysis technique at it without really considering what the data are telling us, or whether the particular analysis technique we are using is the most appropriate for our purpose. In this chapter and the next we will look more critically at regression analysis and consider the times when the results it is giving us may be misleading.

This chapter is very long; in fact it is the longest chapter in the book. We could cover the assumptions in regression analysis very quickly – many texts cover them in a few pages. In this text we have tried not just to tell you what the assumptions of regression analysis are (if we wanted to do that we could write a list), but to give you more of an idea of why those assumptions need to be made and what happens when they are violated. To do this we will return to the structure of Chapter 1. In that chapter we discussed the mean as a least squares model to represent a dataset. We then extended the idea to show how the mean can be treated as a special case of a regression equation. In this section we will do a very similar thing. We first introduce the mean and the assumptions that are made when using the mean as a model to represent a dataset. Because we all understand the mean, it is relatively simple to understand what assumptions need to be made when using the mean, and what might occur when those assumptions are violated. We then develop the assumptions to show how they relate to the fitting of regression lines.

4.2 Assumptions about measures

4.2.1 Levels of measurement

Since research in psychology is concerned with the collection of data, and data are the result of measurements, it is important to have an understanding of the types of scales that are used when making such measurements. The type of measuring scale that is employed will determine the type of statistical tests that can be carried out on the data. A very important series of articles by Stevens (1946, 1951) introduced the idea that there are four types of scales or levels of measurement that are used in psychology experiments. These are referred to as *nominal, ordinal, interval* and *ratio* scales. We will describe them briefly here, but for a more detailed discussion we suggest that you look at Stevens (1951) or Nunnally and Bernstein (1994).

4.2.1.1 Nominal scales

Data that are measured on a nominal scale are often also known as categorical data. With this type of scale observations are sorted into discrete, mutually exclusive categories. Keeping a record of the sex of each participant in an experiment is an example of collecting nominal data. With this example, males could be put into a category labelled '1', females could be put into a category labelled '2', and 'don't knows' into a category labelled '3'. The nominal scale in this example ranges from 1 to 3. The actual numbers under which people or observations are classified are quite arbitrary. That is, the nominal number 2 in this example is not greater or better than 1, it is merely *different*. As this is the case calculating the mean is *not* appropriate for nominal data. Categorising people according to favourite food, experimental group, occupation or eye colour are other examples of collecting nominal data.

4.2.1.2 Ordinal scales

Ordinal data are collected when objects or individuals are put into a rank order. Ordinal data are also sometimes called ordered categorical data. For example, an experimenter who asks you to rank five TV programmes into an order of preference — so that 1 signifies your most favoured and 5 signifies your least favoured — would be collecting ordinal data. The ordinal scale in this example ranges from 1 to 5. If you have a go at this you might notice that the differences in preference between TV programmes are not always equal. There might be little difference in preference between your favourite programme (rank order position 1) and your second favourite (rank order position 2), whilst there might be a big difference in preference between your third favourite (rank order position 3) and your second favourite. Another example of ordinal data could be the order in which runners in a race cross the finish line. The runner who crosses the line first could be two minutes ahead of the runner who crosses the line second, whilst the runner who crosses the line in second position could be less than a second ahead of the runner in third position. So the positions give us information about the order in which the runners finished but not about the gaps between them. In other words, it can be said that ordinal scales do not necessarily have *equal intervals*. As with nominal scales, therefore, usually you should not add them or perform other mathematical operations on them. Whoever said 'a miss is as good as a mile' might have been talking about an ordinal measurement scale.

4.2.1.3 Interval scales

Interval data represent a much 'higher' level of measurement than ordinal data. In order to understand interval data it is perhaps best to consider first an example of an interval scale, such as UK and US shoe sizes. The intervals between adjacent values in the shoe size scale are equal. As this is true, numbers from this scale can be added and subtracted (and therefore we can calculate a mean value). For example, the difference in size between a size 6 foot and a size 9 foot is the same as the difference in size between a size 8 foot and a size 11 foot. We can also say that the size difference between a size 2 foot and a size 6 foot is twice as large as the difference between a size 6 foot and a size 8 foot. However, interval scales do not have a true zero point. You cannot have size 0 feet. In addition, you cannot multiply the values on an interval scale (and expect to get something meaningful out). A person with size 12 feet does not have feet that are twice the size of someone who is a size 6.

Some psychologists consider IQ (intelligence quotient) scores to be interval data (more conservative psychologists would argue that they are ordinal data). This means that a person with an IQ of 120 (high) would be considered to be 20 points more intelligent than a person with an IQ of 100, and a person with an IQ of 140 (very high) is the same amount more intelligent than the person with an IQ of 120. However, a person with an IQ of 0 cannot be said to have no intelligence; they simply have the lowest IQ that the scale can measure.

4.2.1.4 Ratio scales

A ratio scale has all the properties of the interval scale. The ratio scale, however, *does* have an absolute zero point, and therefore represents the highest level of measurement. The measurement of response times in a psychology experiment is an example of the recording of ratio data. As the ratio scale has a true zero (e.g. a clock starts ticking at 0 seconds) we can determine the ratios of values. For example, a participant in an experiment who took 10 seconds to respond to a stimulus can be said to have taken twice as long as a participant who took 5 seconds to respond (a ratio of 2 to 1).

Table 4.1 contains a summary of the levels of measurement adapted from Stevens (1951).

TABLE 4.1

Scale	Operation	Examples
Nominal	Equal versus not equal	Telephone number
Ordinal	Monotonically increasing	Rank order in class test
Interval	Equality of intervals	Temperature (Celsius)
Ratio	Equality of ratios	Temperature (Kelvin)

Interval and ratio data are sometimes grouped together and referred to as either cardinal data or continuous data.

4.2.2 Conservative interpretation of assumptions

In the context of regression, we make the following (strict) assumptions about the nature of our data:

- The dependent variable should be measured on a continuous (interval or ratio) scale.
- The independent variable(s) should be measured on a continuous scale *or* if the independent variables are measured on categorical scales they can be used, after a little recoding (see Chapter 3).

4.2.3 A more liberal approach

Whether we believe we have satisfied or violated these assumptions all depends on what is meant by 'continuous data'. If 'continuous data' means data that are not measured on a truly interval or ratio scale and that therefore they cannot be used in regression analysis, we are in a bit of a mess – very little data collected in psychological research can be truly defined as continuous. There is, however, a (very) fuzzy line between what can definitely be called 'ordinal' and what can definitely be called 'interval'.

Consider the five foods listed in Table 4.2. If we wanted to know how much people liked each of these foods, we could simply ask them to put the foods in order, with their favourite food first, then their next favourite, and so on. This is shown under the column named 'Ordinal response' and this is definitely an ordinal scale. All we know is that the person who has filled in the first column in the response sheet likes milk more than they like any of the other foods listed. Alternatively, we could ask people to give ratings of each of these foods on a scale from 0 to 10. The result of this is shown in the column labelled 'Rating'. Now we can see that the respondent is very fond of milk, and likes bread only a little less than this. In comparison, they do not like either haggis[1] or apples.

TABLE 4.2

Food	Ordinal response	Rating
Apples	5	0
Milk	1	10
Bread	2	9
Haggis	4	1
Carrots	3	8

The rating scale obviously contains a lot more information than the ordinal (ranking) scale – we know that the respondent does not differentiate a great deal between milk and bread, but does differentiate a great deal between carrots and haggis, which the ordinal scale did not tell us. So is this an interval scale? If we apply the rules in a strict sense then no, the scale is not truly interval. We do not know for sure, for example, that the size of the interval from milk to bread is the same as the size of the interval from haggis to milk, although the gap is 1 in both cases. If it were a true interval scale, we would know that these units were always the same size (a gap of 1 centimetre is *always* a gap of 1 centimetre, whether it is the first or the 992nd centimetre).

Therefore, if we were to apply the assumptions in a strict (or what Abelson (1995) calls a 'stuffy' way), we would almost never be using regression. The question becomes not 'Is my variable measured on an interval scale?' but rather 'Is my scale sufficiently close to an interval scale?' Now we are stuck with another problem – defining how close is close enough.

One thing to consider is the number of categories your respondents are able to use – the more categories, the closer to an interval scale you are likely to be. Use as many categories as you think your respondents will be able to respond to sensibly. Seven is considered to be the optimum number, but people who are used to responding to rating scales can often use more points, while people who are not used to them may prefer to use fewer points.

4.3 Assumptions about data

When we do regression analysis, we make a number of assumptions about the distribution of the dependent variable and the distribution of the residuals (we encountered residuals in Chapter 1). If these assumptions are not satisfied, the conclusions we derive from the analysis will not have a sound basis and will be incorrect. In this section, we will consider in what ways data can violate this assumption and then look at what we can do about problems that do arise. First, we will look at a simple case, the mean, in detail. We will then build up to the more complex cases, those involving residuals in bivariate and multiple regression.

4.3.1 A bit about normal distributions

The normal distribution, also sometimes called a *Gaussian distribution*, is a very important distribution in statistics. Normal distributions when plotted on a frequency distribution form the characteristic symmetrical bell-shaped curve. A frequency distribution that shows a normal distribution is shown in Figure 4.1.

For accurate calculation of standard deviations, and therefore of standard errors, we must have a normal distribution. If the data are not normally distributed the standard errors (and therefore the significance tests) may be inaccurate.

4.4 Univariate distribution checks

In the univariate case (where there is only one independent variable), regression analysis assumes that:

- The residuals are normally distributed.

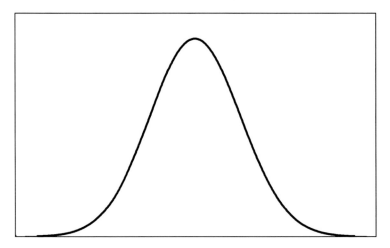

FIGURE 4.1 *A normal distribution*

(Note that we do not make such an assumption about the predictor variable – but more on that later.) Before we consider the case of the independent variable and the residuals, we will go back to thinking about the mean. In Chapter 1 we built a simple model of a dataset – the mean and the standard error of the mean. We were interested not just in the results for the sample of students that we had taken, but also in what the results would be for the whole population of students from which that sample was taken. Recall that when we said that the mean number of books read by students in the class was 2, we wanted a model that described the number of books read by students and the mean provided us with a reasonable model. In addition, we calculated the standard error of the mean, which was 0.23. These two pieces of information give us an indication of the mean number of books read by the whole population (all of the students) and an indication of how accurate this estimate is. For the calculation of the mean and standard error to be any use to us we must be able to assume that the data follow the pattern of a normal distribution. There are two related ways in which the data can fail to follow the pattern of a normal distribution:

1. The data contain a small number of scores that are extremely large or extremely small compared to the rest; these scores are termed *outliers*.
2. The shape of the whole distribution fails to resemble the bell-shaped curve which is characteristic of the normal distribution.

We will examine the effects that these two failures to follow the normal distribution pattern have on the processing of our data.

4.4.1 Outliers and the mean

Outliers are aberrant scores that lie outside the usual range of scores we would expect for a particular variable. For example, an annual salary of £250 000 would be considered an outlier in a distribution of lecturers' salaries. If there are outliers in a set of data, the mean may not be an appropriate way of representing these data — it will not therefore be a good model. Here is an example that shows why not: the salaries of a group of members of staff in a university department are shown in Table 4.3.

TABLE 4.3

Person	Salary (£000's)
Technician 1	10.4
Technician 2	11.3
Technician 3	12.9
Lecturer 1	13.2
Lecturer 2	14.6
Lecturer 3	15.8
Lecturer 4	15.5
Lecturer 5	16.1
Lecturer 6	17.0
Senior Lecturer 1	22.7
Senior Lecturer 2	23.5
Senior Lecturer 3	24.6
Head of Department	28.9

If we calculate the mean amount earned by members of staff in that department, we find the mean is £17 400 per year. This model is a reasonable description of how much the members of staff earn. We can tell it is a good model, because if we had no idea how much someone earned, but guessed it would be £17 400, we probably would not be too far off — in the same way that you were not too far off when you guessed how tall we were in Chapter 1. What happens when one of the staff members in that department gets a large pay rise? For example, one of the members of staff may write a particularly successful book, say on regression analysis, and receive a very dramatic pay increase. The table may then look like that in Table 4.4.

TABLE 4.4

Person	Salary (£000's)
Technician 1	10.4
Technician 2	11.3
Technician 3	12.9
Lecturer 1	13.2
Lecturer 2	14.6
Lecturer 3	15.8
Lecturer 4	15.5
Lecturer 5	250.6
Lecturer 6	17.0
Senior Lecturer 1	22.7
Senior Lecturer 2	23.5
Senior Lecturer 3	24.6
Head of Department	28.9

The mean salary now becomes £35 500. This is no longer a representative model — it does not describe the amount of money that anyone earns, being much less than the large salary and much more than the general run of staff salaries. If we guessed that someone in the department earned £35 500 we would not be close. This is not a good model. In technical terms, the figure for the salary of Lecturer 5 has become an *outlier*; it is outside the range that we would expect if we had a normal distribution. The inclusion of the value of £250 600 has had a very large effect when the new mean score was calculated.

4.4.2 Normal distribution

If there are no outliers occurring in the data, the distribution may still deviate from normality. From Figure 4.1 you can see that the normal distribution is symmetrical and approximately bell shaped.

The distribution can deviate in two ways:

1. The distribution can be non-symmetrical. This means that one tail of the distribution is longer than the other tail. We describe a non-symmetrical distribution as *skewed*.
2. If the distribution of the data is too flat or too peaked, that is the tails are too short or too long, the distribution is described as being *kurtosed*.

4.4.2.1 Skew

Skew occurs if the scores are not symmetrically distributed. Skew can occur for a number of reasons but it most commonly occurs when there is some sort of *floor effect* or *ceiling effect*. A floor means that the data have a minimum value. For example, in a reaction time experiment all the scores have to be greater than 0, as it is not possible to react in less than zero seconds. But some of the reaction times could be very long if a participant hesitated or was momentarily distracted. This would cause positive skew (scores bunching together at the lower end of the range). A ceiling effect occurs when it is not possible to score above a certain upper limit. For example, in a poorly designed test with 10 easy questions it may happen that all the scores are bunched at the top because most people got most of the questions correct. This would cause negative skew (scores bunched together at the top of the scale).

A great many students (not to mention researchers) have difficulty remembering what is positive skew and what is negative skew. To remember which is positive and which is negative, it is helpful to see (Figure 4.2) that the negatively skewed distribution starts off flat, in the same way as a minus sign, whereas the positively skewed distribution starts off going up, as in the upper stroke of a plus sign.

4.4.2.2 Kurtosis

Kurtosis causes fewer problems in the estimation of regression models than skew. This is fortunate because kurtosis is also a much stickier statistical issue than skew, as noted by DeCarlo (1997) who examined issues around kurtosis

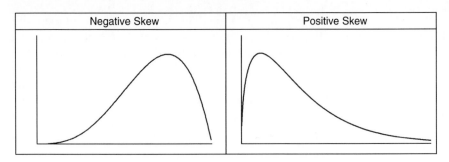

FIGURE 4.2 *Skewed distributions*

and found that a number of major textbooks even became confused and defined kurtosis incorrectly.

If the distribution is symmetrical, but does not have the characteristic bell shape, it is said to exhibit kurtosis. Kurtosis should be thought about in terms of the tails of the distribution. A distribution can be too 'flat', as if it has been squashed; in other words, the tails are too heavy. If the curve is too flat, the distribution is described as being *negatively kurtosed* or *platykurtosed* (derived from the French word *plat*, meaning flat – also the derivation of plate). Alternatively, if the distribution is too 'peaked', as if it has been pulled up at the centre and there are not enough values near the tails, then the distribution is described as being *positively kurtosed*, or *leptokurtosed* (derived from the Greek word *leptos*, meaning small, or slender).[2] Figure 4.3 shows three distributions: normal, leptokurtosed and platykurtosed.

Table 4.5 contains a set of values of x from 1 to 11 in the first column. In the three columns to the right are frequency counts for distributions with three different levels of kurtosis. If you compare the negative kurtosis with the

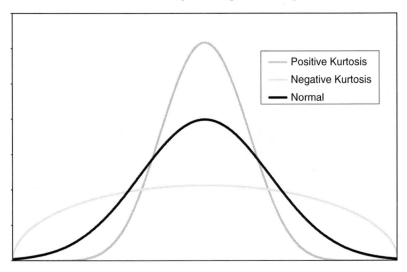

FIGURE 4.3 *Normal and kurtosed distributions*

TABLE 4.5 *Frequency table showing normal, negatively kurtosed and positively kurtosed distributions*

x	Negative kurtosis	Normal	Positive kurtosis
1	2	2	2
2	14	4	3
3	30	12	4
4	38	24	10
5	41	36	20
6	42	42	42
7	41	36	20
8	38	24	10
9	30	12	4
10	14	4	3
11	2	2	2

normal, you will see that there are higher frequencies towards the ends of the distributions in the negative than in the normal. Similarly, there are a larger number of higher frequencies in the tails of the normal distribution than there are in the positively kurtosed distribution.

4.4.3 Detecting and dealing with non-normality

In this section, we will first consider the methods for detecting non-normality in a distribution that has a single variable. There are two types of method: graphical methods and numerical methods. We will then move on to the multivariate case and consider the issue of the multivariate normal distribution. Finally, we will examine what can be done about non-normality.

4.4.3.1 Graphical methods

In this section, we aim to give a brief survey of some of the methods that are used for detecting non-normality. Many of these methods are just different ways of graphically representing the same data, and show the same things. For a much more thorough examination of methods of graphically exploring your data you should have a look at Tukey (1977).

4.4.3.1.1 Histograms

A histogram is the easiest way of defining a normal distribution. Distributions are usually defined by the shape that they form when they are drawn on a histogram. A histogram is a plot that shows the values along the x-axis (the horizontal axis) and the number of values that obtained that score on the y-axis (the vertical axis). A histogram can be drawn as either a bar chart or a line chart (although then, strictly speaking, it should be called a frequency distribution). Some authors will claim that one is more or less appropriate than another for different data types, but we feel that the most important criterion is to ensure that your data are displayed clearly.

Detecting skew, kurtosis and outliers is easy using a histogram. If the data appear to be bell shaped and symmetrical then the distribution is probably

68 APPLYING REGRESSION AND CORRELATION

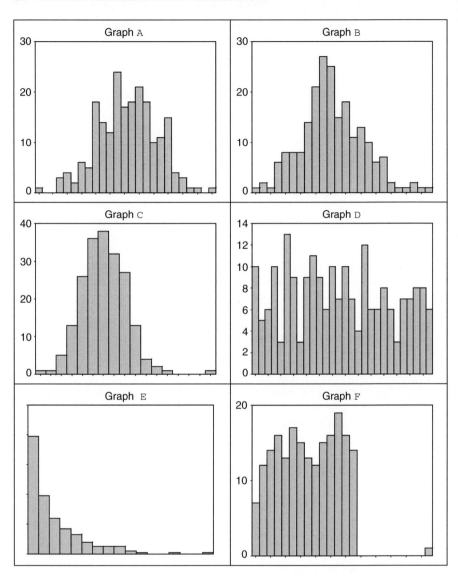

FIGURE 4.4 *Graphs that might show departures from the normal distribution*

approximately normal. If the shape of the distribution is not bell shaped and symmetrical then your data cannot be assumed to be normally distributed.

The graphs in Figure 4.4 show examples in the form of histograms of normal distributions and distributions that vary in some degree from the normal distribution. Before reading ahead, have a look at the graphs and try to determine for yourself whether they show an approximately normal distribution, and if they do not, ask yourself in what way they depart from the normal distribution (each of them is from a sample of size $N = 200$).

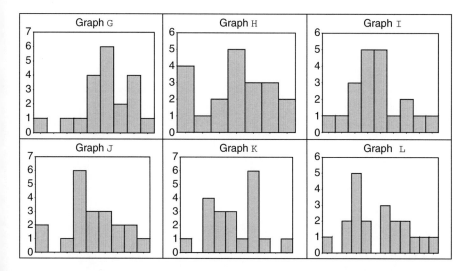

FIGURE 4.5 *Histograms* ($N = 20$)

Graph A seems to be approximately symmetrical, and follow the correct bell-shaped curve. If you said that graph A was normally distributed, you were right. Graph B is also normally distributed, or at least not too far from it. Graph C has the characteristic shape, but does not have the symmetry we would expect of a normal distribution. There is an outlier on the right-hand side of the graph, which is causing the distribution to deviate from normality. Graph D has the symmetry that we would expect of a normal distribution but is negatively kurtosed. The tails of the distribution are too heavy and the shape of the distribution does not form the characteristic bell shape that we would expect. Graph E appears at first to have two outliers on the right-hand side. If the graph is examined more closely, it can be seen that the graph would not become normal if these two outliers were removed. Rather this distribution is actually positively skewed. Graph F is not normally distributed for two reasons. First, there is an outlier on the right-hand side of the graph. However, even if this outlier were removed, the distribution would still not form the bell shape that we desire − it would be negatively kurtosed.

So, if histograms are the standard way of defining distributions, why are we going to bother looking at any other methods? The graphs shown in Figure 4.5 all have a sample size of 20. Before reading ahead try to decide which of them are randomly sampled from a normal distribution, and which are not.

All of the graphs shown in Figure 4.5 are randomly sampled from populations in which the distribution is normal. Because of the small sample size, it appears that the shape of the distribution has deviated from the normal distribution, when in fact it has not. Because it is easy to be fooled by histograms, especially when samples are small, other methods of checking for normal distribution may be more appropriate.

70 APPLYING REGRESSION AND CORRELATION

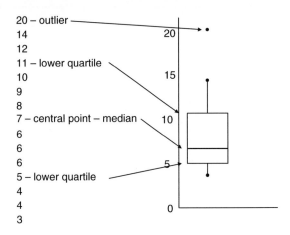

FIGURE 4.6 *Construction of a boxplot*

4.4.3.1.2 Boxplot

A boxplot, also called a box-and-whisker plot (developed by Tukey, 1977), is another graphical way of looking at the data. Figure 4.6 shows how a box-and-whisker plot is constructed using the following data: 3, 4, 4, 5, 6, 6, 6, 7, 8, 9, 10, 11, 12, 14, 20.

The first stage in drawing a boxplot is to find the median. When the data points are ranked from highest to lowest, the middle point represents the median. The median is represented as a thick horizontal line on the chart. The next stage is to mark the quartiles; in the same way as the median represents the half-way point the quartiles represent the one-quarter- and three-quarter-way points. These are drawn as a box around the median. The whiskers extend from the edge of the box to the highest and lowest points *unless* those points are more than ±1.5 times further from the central line than the edge of the box. Points that exceed this distance are drawn as points and represent outliers.

Because the boxplot summarises the data, and removes some of the 'lumpiness', it is easier to use a boxplot than a histogram to see when a distribution is deviating from normality. Figure 4.7 shows the same data that were shown in Figure 4.5 as histograms, but this time displayed as boxplots. The same characteristics can be seen in these as in the previous graphs. A and B are normally distributed as the shapes are symmetrical around the median and the whiskers are longer than the boxes. There are some outliers, but these are not extreme, as they are not far from the ends of the whiskers – and this is acceptable in a sample of size 200. The dataset C appears to follow the same pattern and be approximately normal, except for the outlier that could also be seen in the histogram of the same data. Distribution D is symmetrical, and therefore satisfies our first criterion, but the whiskers do not extend as far as they would if the distribution were normal. Distribution E has a number of extreme outliers and we can also see that it is skewed, from the way that the

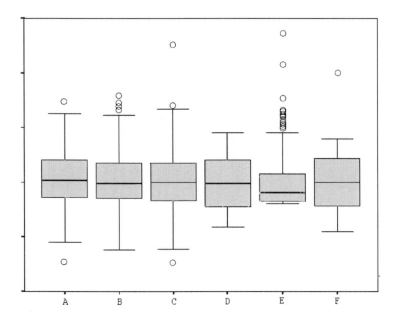

FIGURE 4.7 *Boxplots*

median line is not central within the box, and the whiskers are very different lengths. Finally F can be seen to have an outlier, and, like D, the whiskers are too short, relative to the box length, for the distribution to be normal.

As we said earlier, an advantage that the boxplot has over the histogram is that for a smaller sample size random deviations from normality can make a histogram appear non-normal, but these few deviations do not have the same effect on the boxplot. This is demonstrated in Figure 4.8, where the data that were used to produce the histograms shown in Figure 4.5 are used. In these charts the distributions, although not perfectly normal, do not seem to differ from normality to any large degree.

4.4.3.1.3 Probability plots

A probability plot (or P–P plot) is a more mathematical method of comparing data with a normal distribution. We know (or it is possible to find, either by calculating or by looking up a table) the types of scores that we would expect to get if our data were normally distributed. We can use this information to compare our dataset with an imaginary dataset that would be found in an 'ideal' normal distribution with the same mean and standard deviation. If the distribution we are interested in matches the normal distribution fairly well, we can conclude that our data are normally distributed. If we find a reasonable match, the points of the probability plot that we generate will lie in a straight line along the diagonal from the bottom left to the top right. If the distribution differs from normality, then the points will lie further from the diagonal.

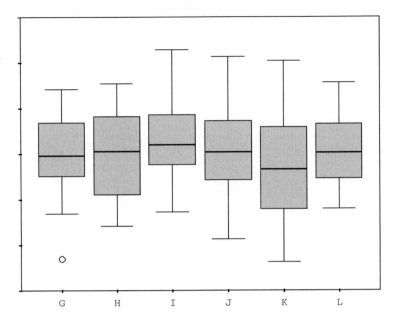

FIGURE 4.8

Figure 4.9 shows the set of distributions that we are becoming familiar with, displayed as P–P plots. We can see that graphs A and B show that the points are falling fairly close to the diagonal line, indicating that the distribution is normal. Graph C also seems to lie close to the line, again showing that the distribution is normal, but we know from previous analyses that it is not normal – there is an outlier in this dataset. This example shows us that P–P plots are not very good at detecting a small number of outliers in a large dataset. Graphs D, E and F show what P–P plots are good at, though, and this is detecting trends away from general normality. The curve of the line in graph D deviates from the straight line in a symmetrical fashion, tending slightly towards an S-shape. The fact that the curve is symmetrical (if rotated through 180°) tells us that the distribution is symmetrical, and therefore not skewed, but the fact that it is not straight tells us that the distribution is non-normal. In graph E the curve deviates from a straight line and is also non-symmetrical, meaning that the distribution is skewed. Finally in graph F we can see the characteristic S-shape of the curve, but we can also see the outlier where the line jumps to the end very suddenly.

4.5 Calculation-based methods

Graphical methods for determining whether a variable is normally distributed involve some calculation, particularly in the P–P plots, but there are some methods to detect outliers employing calculation only: these we call the *calculation-based methods*.

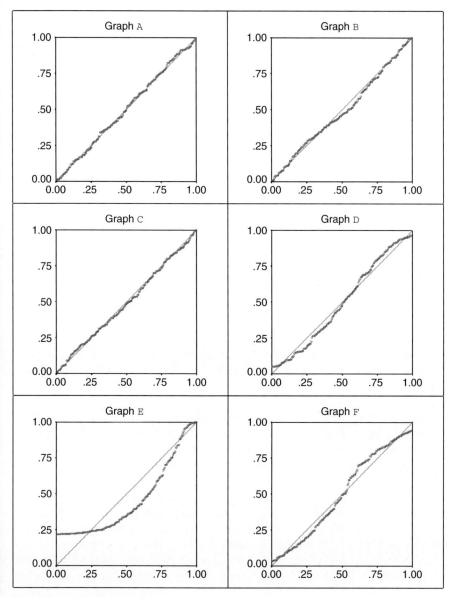

FIGURE 4.9 P–P plots

4.5.1 Skew and kurtosis

It is possible to get numerical values for the degree of skew and kurtosis of a variable, and these are called, respectively, skew and kurtosis. The skew and kurtosis of a variable that is normally distributed will both have the value 0. Variables that depart from normality will be indicated by values above or below zero.

There are two slightly different methods for calculating the skew and kurtosis statistics, which were developed by the rival statisticians Fisher and Pearson.[3] Which one you use depends on which statistical package you use. Fisher's technique is used by SPSS, SAS and MS Excel, and Systat, while Pearson's technique is used by STATA.

In addition to the values for the skew and kurtosis, most statistical packages calculate a standard error of the skew and kurtosis. These values can be used to help to determine whether the skew and kurtosis differ *significantly* from what might reasonably be expected in a normally distributed population: if the value of skew or kurtosis (ignoring any minus sign) is greater than twice the standard error, then the distribution significantly differs from a normal distribution. You should be aware that there is a problem with using these measures of standard error, which is part of a much larger debate around significance testing in general (see e.g. Cohen, 1994; Chow, 1996). This problem is that we are not really interested in whether the distribution is *significantly* different from a normal distribution (using significant in the technical sense of the word). We are really interested in knowing whether the distribution is sufficiently skewed that it *matters*. We usually shy away from rules of thumb, but we cautiously suggest that if your skewness statistic is less than 1.0, there should be little problem. If the skewness is greater than 1.0, but less than 2.0, you should be aware that it might be having an effect on your parameter estimates, but that it is probably OK. Finally, if the skewness statistic is greater than 2.0 you should begin to be concerned. Each of these values depends upon your sample size – the larger your sample size, the less departures from normality matter.

Table 4.6 shows the skew and kurtosis statistics (calculated by SPSS) of the variables A, B, C, D, E and F that we have used in this chapter. The skew and kurtosis statistics for A and B are both close to zero, and are not greater than twice their standard error (ignoring minus signs). We can happily say (as we have been saying all along) that the variables A and B are both normally distributed. Variable C has a moderate degree of skew (0.454); this is significant, as 0.454 is greater than 2 × 0.172, but probably not high enough to concern us. The value for the kurtosis of variable C is both significant and high enough, at 1.8, to warrant concern. Variable D is not skewed, as we saw from the graphical methods, but as we also saw, it is not sufficiently 'peaked' to be normally distributed – the kurtosis figure is both high and significant.

TABLE 4.6 *Skew and kurtosis values, and their associated standard errors*

	A	B	C	D	E	F
Skewness	−0.120	0.271	0.454	0.117	2.106	0.171
SE skew	0.172	0.172	0.172	0.172	0.172	0.172
Kurtosis	−0.084	0.265	1.885	−1.081	5.750	−0.210
SE kurt.	0.342	0.342	0.342	0.342	0.342	0.342

Variable E causes concern from the point of view of both skew and kurtosis – both are high and significant. Finally, variable F, from these statistics, seems to be close enough to a normal distribution not to cause us concern.

The skewness and kurtosis statistics (along with their associated standard errors) for the six smaller datasets that we previously examined in Figure 4.5 are shown Table 4.7. These statistics show that the distribution do depart from normality, though they do not depart to a significant degree.

TABLE 4.7 Skew and kurtosis statistics for variables G to K

	G	H	I	J	K	K
Skewness	−0.569	−0.205	0.493	0.046	0.059	0.265
SE	0.512	0.512	0.512	0.512	0.512	0.512
Kurtosis	0.922	−1.005	0.045	−0.382	−0.549	−0.856
SE	0.992	0.992	0.992	0.992	0.992	0.992

4.5.2 Outliers

The skew and kurtosis statistics are very good at detecting general deviations from the normal distribution, but they are not good at detecting outliers. There are many methods for determining whether a case should be considered an outlier when we are using regression analysis. We will be looking in the first instance at the univariate case, and then we will consider the more complex cases that arise in a multiple regression situation.

4.5.2.1 Calculating Z-scores

If we have a normal distribution, we can find out (by using tables such as the one in Appendix 3) what proportion of data points we expect to find at different distances from the mean. We know from the tables that (approximately) 68% of scores will lie within one standard deviation from the mean, 95% of scores will lie within two standard deviations of the mean, and 99% of scores will lie within three standard deviations of the mean. We can convert a score into what is called a standardised score, or a z-score, by calculating the number of standard deviations from the mean that any score lies. To do this we subtract the mean and divide the score by the standard deviation, as shown in the following equation (most statistical packages have a function that will automatically standardise a variable):

$$z(x) = \frac{x - \bar{x}}{sd}$$

It follows that if we have a dataset that contains the z-scores of a normally distributed measure (e.g. height) of a sample of 100 people, we would expect to find several scores whose z-value was greater than two and approximately one that had a z-score greater than three standard deviations. We would, however, be surprised to find any values greater than four standard deviations from the mean, as a score this extreme occurs, on average, approximately only once in 7500 cases.

Table 4.8 shows the standardised scores of the salary data that we first encountered in Tables 4.3 and 4.4. You can see that the very high salary has a z-score of 3.3. We would expect to find one value with a z-score of approximately 3 in a dataset of 100, but the fact that we have one in a dataset of only 13 suggests that that datum should be considered an outlier.

TABLE 4.8

No outlier		Outlier	
Raw score	Standardised score	Raw score	Standardised score
10.4	−1.23	10.4	−0.39
11.3	−1.07	11.3	−0.37
12.9	−0.79	12.9	−0.35
13.2	−0.74	13.2	−0.34
14.6	−0.50	14.6	−0.32
15.8	−0.28	15.8	−0.30
15.5	−0.34	15.5	−0.31
16.1	−0.23	250.6	3.32
17.0	−0.07	17.0	−0.28
22.7	0.93	22.7	−0.20
23.5	1.07	23.5	−0.18
24.6	1.26	24.6	−0.17
28.9	2.01	28.9	−0.10

Using standardised scores to detect outliers is not a wholly satisfactory method. We used the mean to calculate the z-score for each person, but we have already said that the mean is not useful in this dataset, because there is an outlier. A better method is to use the *deleted z-score*. The deleted z-score is the z-score of that datum, but instead of using the mean and standard deviation of all of the data points, we use the mean and standard deviation of all of the data points *except the one that we are interested in*. To calculate the deleted z-score, the first case is deleted, the mean and standard deviation for the rest of the data are calculated, and this mean and standard deviation are then used to calculate the z-score for the deleted case. The process is then repeated, removing each case, one at a time.

The standard z-scores and the deleted z-scores are shown in Table 4.9. You can see that the extreme values have all increased in both sets of data, but the outlier score has increased much more than any other score.

4.5.2.2 Influence statistics

Influence statistics are rarely used in the univariate case to determine if a data point should be considered an outlier, but because influence statistics are used in regression, we shall introduce them here. The aim when using an influence

TABLE 4.9

No outlier			Outlier		
Raw score	z-score	Deleted z-score	Raw score	z-score	Deleted z-score
10.4	−1.23	−1.13	10.40	−0.39	−0.09
11.3	−1.07	−0.93	11.30	−0.37	−0.08
12.9	−0.79	−0.61	12.90	−0.35	−0.05
13.2	−0.74	−0.55	13.20	−0.34	−0.05
14.6	−0.50	−0.29	14.60	−0.32	−0.03
15.8	−0.28	−0.07	15.80	−0.30	−0.01
15.5	−0.34	−0.12	15.50	−0.31	−0.01
16.1	−0.23	−0.01	250.60	3.32	42.54
17.0	−0.07	0.15	17.00	−0.28	0.01
22.7	0.93	1.23	22.70	−0.20	0.10
23.5	1.07	1.40	23.50	−0.18	0.12
24.6	1.26	1.65	24.60	−0.17	0.13
28.9	2.01	2.89	28.90	−0.10	0.20

statistic is to see how much each individual data point influences the model parameters, even if the only parameter of interest is the mean. If the score of one person can dramatically change the parameter estimate, that person's score is probably having an undue influence on the results (hence the name influence statistics).

An influence statistic is calculated by:

1. Calculating the parameter estimate for all of the variables (e.g. the mean).
2. Recalculating the parameter estimate with the one data point excluded.
3. Calculating the difference between the results of 1 and 2.

(It is probably not necessary at this point to worry about the calculation of this statistic; when they are actually used in regression analyses most packages will calculate them automatically.)

Table 4.10 shows the change statistics for the data with the outlier and for the data without the outlier. It is easy to see that in the first part of the table, the difference scores are all small and, more importantly, they are similar. In the second part of the table, the difference scores are larger and all similar except for the difference score of the eighth data point number, which is much higher, indicating that this data point is an outlier.

4.5.2.3 Standardised influence statistics

In Chapter 1, we encountered the problem of the variables having different and arbitrary scales, which meant that scale-dependent calculations were influenced in a way that made them difficult to interpret. We have used a scale that measures pay in thousands of pounds, but if we had used a scale that measured pay in single pounds instead, all the values would have been a thousand times higher. If we had used francs, dollars or euros, the numbers would be different again. In Chapter 1, when we encountered the problem of scale, we used standardised scores. Standardised scores convert the mean to

TABLE 4.10

	No outlier			Outlier	
Raw score	Mean if case excluded	Difference	Raw score	Mean if case excluded	Difference
10.40	18.01	−7.61	10.40	37.55	−27.15
11.30	17.93	−6.63	11.30	37.48	−26.18
12.90	17.80	−4.90	12.90	37.34	−24.44
13.20	17.78	−4.58	13.20	37.32	−24.12
14.60	17.66	−3.06	14.60	37.20	−22.60
15.80	17.56	−1.76	15.80	37.10	−21.30
15.50	17.58	−2.08	15.50	37.13	−21.63
16.10	17.53	−1.43	**250.60**	**17.53**	**233.07**
17.00	17.46	−0.46	17.00	37.00	−20.00
22.70	16.98	5.72	22.70	36.53	−13.83
23.50	16.92	6.58	23.50	36.46	−12.96
24.60	16.83	7.78	24.60	36.37	−11.77
28.90	16.47	12.43	28.90	36.01	−7.11

zero and the standard deviation to one, and they make the expected change statistic much easier to interpret.

Table 4.11 shows for each value the raw scores, the standardised scores and the standardised change in the mean. You can see that the change in the mean associated with the outlier is much higher than for any other score.

TABLE 4.11

	No outlier			Outlier	
Raw score	Standardised score	Change in mean	Raw score	Standardised score	Change in mean
10.40	−1.23	−0.10	10.40	−0.39	−0.03
11.30	−1.07	−0.09	11.30	−0.37	−0.03
12.90	−0.79	−0.07	12.90	−0.35	−0.03
13.20	−0.74	−0.06	13.20	−0.34	−0.03
14.60	−0.50	−0.04	14.60	−0.32	−0.03
15.80	−0.28	−0.02	15.80	−0.30	−0.03
15.50	−0.34	−0.03	15.50	−0.31	−0.03
16.10	−0.23	−0.02	250.60	3.32	0.28
17.00	−0.07	−0.01	17.00	−0.28	−0.02
22.70	0.93	0.08	22.70	−0.20	−0.02
23.50	1.07	0.09	23.50	−0.18	−0.02
24.60	1.26	0.10	24.60	−0.17	−0.01
28.90	2.01	0.17	28.90	−0.10	−0.01

4.6 Dealing with outliers, skew and kurtosis

If the distribution is not normal, either because of outliers or because of skew or kurtosis, least squares estimates and their standard errors (and remember, the mean is an example of a least squares estimate) will be inaccurate. We now have a number of choices for dealing with a non-normal distribution: we could use a technique other than a least squares technique, for example we could use what is often called a non-parametric test (sometimes called a distribution-free test). Unfortunately, whilst simpler analyses often do have a straightforward non-parametric equivalent, regression does not. We could learn from our experience, refine our measures, go and collect some more data, and hope that they would be normally distributed. This would be something of a waste though – we have probably spent a great deal of time and money collecting those data. Or we could try to do something about the data to try to make them suitable for analysis.

4.6.1 Dealing with outliers

Deciding whether a data point is an outlier and deciding what action to take is much more of an art than a refined science. The first thing to do is to try to determine *why* the outlier has occurred.

If it is possible that the outlier occurred because of faulty measurement equipment, or an error while entering data into the computer, you should wherever possible go back and check, and find the correct value – it is often easiest to tell that this has happened when a value could not have been achieved. If a participant has scored 77 on a rating scale that went from 1 to 7, it is safe to say that a data entry error has occurred. If you cannot find the correct value you may simply want to delete the data point and carry on without it.

If the outlier or extreme point has occurred because of a true, properly measured data point, then you need to look at both your theory and your measuring scales to see if they are appropriate. In the 'salaries' example that we have been considering, we have made an error in the way that we have measured salary. We should return to our data collection, and make it clear that when we say salary, we are interested in salary paid by the university, excluding money from other sources (book sales, lottery wins, inheritance, fraud). If we restate what we mean by salary in this way, making it more closely related to the theoretical construct that we are actually interested in, we will probably find that the outlier disappears, and our data become more useful.[4]

A common occurrence is to find outliers in measures of reaction time. If they occur, you need to consider why they have occurred. We are trying to model a psychological process: if the outliers are part of that normally functioning psychological process, then they should be included in the model. If they are outside that psychological process, the participant was bored, yawned, had an itch, etc., then they should be excluded.

If the outlier is still there after you have checked your sources of data and reassured yourself that your theory is the appropriate one, you are left with two options, neither of which is wholly satisfactory. The first option is to carry out the analysis with the outlier and be aware that it may be having an undue influence on your parameter estimates. The second option is to delete the outlier, and analyse the data with the outlier excluded. Sometimes, when one outlier is removed and the data are rechecked, another case appears as an outlier, so it is removed, until a large number (too many?) of cases are removed from the analysis. The decision as to the inclusion or removal of outliers is part of the art that you are learning: do you compromise by modelling all of the data badly (leaving outliers in) or do you compromise by modelling most of the data well (excluding outliers)?

In conclusion, we quote from Pedhazur and Schmelkin (1991) who suggest avoiding the compromise by analysing the data twice:

> The onus of interpreting what the outliers and/or influential observations mean in a given set of data and the decision about what to do is, needless to say, on the researcher.... Whatever the reasons for the rejection of observations, the researcher owes the reader a complete reporting of criteria used for the designation of outliers and influential observations, what has been done about them, and why. In addition, generally speaking, major aspects of results of analyses with and without the outliers and/or the influential observations should be reported. (p. 408)

4.6.2 Effects of univariate skew and kurtosis

4.6.2.1 Effects of skew

Because the mean is a least squares estimator of central tendency, points that are further away from the mean will have a greater influence on the results of calculations than nearer ones, and therefore the mean will prove to be a biased estimator — that is, it will not accurately reflect the population. In a negatively skewed distribution, the mean will be biased downward; in a positively skewed distribution, it will be biased upward.

Because measures tend to have lower limits but no upper limits, positive skew is more common than negative skew. Consider the following measures:

Earned income: It is not possible to earn less than no money;[5] therefore the amount of money people earn must start at zero. It is, however, possible to earn *lots* of money so there is almost no upper limit. In the previous example, we considered one department of a university, but if we considered the whole university, we might find a few people who earn very little, for example part-time staff or postgraduate students. Most staff — lecturers, administrators, researchers — will earn something near the mean amount of money, a smaller number of heads of schools and large departments will earn more than the average, three or four very senior managers will earn substantially more, and finally the one person in charge will probably earn the most. The distribution of salary would be positively skewed.

Time: Experiments in psychology, quality control tests in manufacturing, medical tests of drug success, and marketing studies may all measure the time taken for an event to happen. In psychology, we may be interested in the amount of time a participant takes to reach a decision. In manufacturing, we may be interested in how long components made using different processes last before they fail. In medicine, we may be interested in seeing how long patients live for on different drug or treatment regimes.[6] Finally, in marketing we may want to know how long it takes to sell a car advertised in different ways. All of these measures share a measurement of time. This measure of time cannot be less than zero, but could be very large (i.e. a long time).

If we measure reaction time, usually people are fairly quick, but sometimes they miss the button, have an itch, yawn, or were simply not concentrating, and so take a long time to react. Some components of pieces of equipment fail as soon as you get them home, some fail just after the guarantee expires, and some give years of faithful service. All of these measures of time share the fact that the data will be positively skewed because there are no negative or zero values, and so any of these distributions will show positive skew.

Depression: Measures of depression are very commonly used in applied psychology and psychological research. It is usually found that most people are not depressed, or are only very slightly depressed, whereas a few people are more depressed, and a smaller number of people are very depressed indeed. Again, if we plotted a histogram of the distribution of degrees of depression, we would find that the data were positively skewed.

When the data are kurtosed, the mean value of a variable remains the same, but the standard errors of the mean can become too small or too large. The calculations we used for the standard error only give the correct value when the distribution is normal. The distortion that occurs in a non-normal distribution will have one of two effects. First, it may have the effect of making us believe that our estimates are less accurate than they really are – the standard errors will be too large. If the standard errors are too large, the type II error rate will be inflated and, as a result, we will be less likely to find significant effects. If the standard errors are too small, the type I error rate will become inflated and, as a result, we will think our estimates are more accurate than they actually are; we will believe that we have found a difference, or an effect, when in fact we have not.

In the more complex case of regression analysis, kurtosis can affect the parameter estimates as well as the standard errors, but the effects are usually small unless the kurtosis is severe.

4.6.2.2 Dealing with skew and kurtosis: transformations

Transformations are a way of taking data that are not normally distributed, and transforming them in some way to make them fit the normal distribution more closely. We will introduce the idea of transformations using an example.

Table 4.12 shows a set of values derived from a hypothetical experiment on reaction time. The times were measured as the time taken, in seconds, to reach a decision as to whether a word was spelt correctly. The values are mainly in the region of 1 to 3 seconds; however, there are a few values of less than 1 second, and several values longer than 1 second. The longest is 10.79 seconds.

TABLE 4.12 *Reaction time data (dataset 4.1)*

0.92	10.79	6.45	1.24	0.91
2.79	2.38	1.04	4.79	1.74
1.35	6.84	3.44	2.02	3.95
7.30	1.67	0.79	1.33	3.77
1.41	0.82	1.43	2.34	0.92
1.39	2.78	1.69	2.45	3.03
0.64	1.67	1.53	2.38	2.18
4.05	0.81	1.76	2.48	2.73
6.51	0.97	3.10	2.36	2.04
7.28	1.79	1.43	0.83	2.57

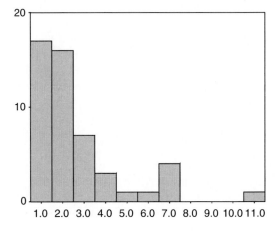

FIGURE 4.10

Figure 4.10 shows a histogram of the data. It can be seen from this that the data are positively skewed.

We can confirm that there is positive skew by calculating the skewness values. Table 4.13 shows some of the descriptive statistics for the variable. The skewness statistic, 1.94, indicates positive skew.

TABLE 4.13

Mean	2.66
Std deviation	2.11
Skewness	1.94
SE	0.34
Kurtosis	4.02
SE	0.66

A transformation is a calculation that is done to all of the values of a variable together. Some transformations will not change the shape of the distribution (multiply all values by 3, or add 5 to all values), other transformations will change the shape of the distribution. A commonly used

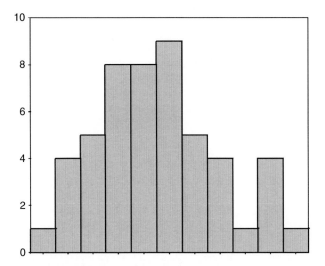

FIGURE 4.11 *Histogram of 'logged' variable*

transformation for removing positive skew is the log transform. (Do you remember learning about logs at school and wondering why you had to bother? We did, and now we know.) To carry out a log transform, we take the logarithm of every value, and use the result (i.e. the logs of the original data) as a variable in our new calculation. Your statistics package or spreadsheet should be able to do this very easily. The histogram of the new 'logged' variable is shown below as Figure 4.11 and the descriptive statistics are given in Table 4.14. Both the graph and the statistics show that the distribution of the transformed data is much more symmetrical. It can be seen in the graph and in the descriptive statistics that the variable now approximates a normal distribution much more closely.

TABLE 4.14 *Distribution statistics for logged variable*

Mean	0.321
Std deviation	0.293
Skewness	0.436
SE	0.337
Kurtosis	−0.316
SE	0.662

When we calculated the mean of the new variable, it was found to be 0.321. This mean value was calculated from the logarithm of the raw scores rather than the raw scores themselves. It is necessary to transform the value back onto the raw score scale. The 'unlogging' or 'antilogging' is done using the exponential (often called EXP) function in a statistics package, or it can be done with a scientific calculator. The exponential of 0.3207 is 2.092, and

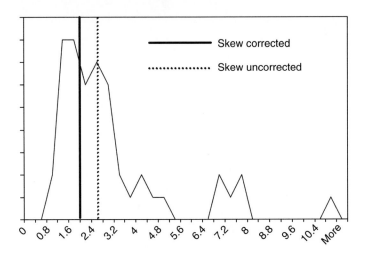

FIGURE 4.12 *Frequency distribution with uncorrected, and log corrected, mean score*

therefore we can treat 2.092 as the corrected mean. The graph in Figure 4.12 shows the distribution of the original data first with the mean calculated without taking the skew into account and second with the mean calculated using the log transform. By using this transformation, we have calculated a value for the mean that is closer to most of the values – it has not been affected by the skew.

Other transformations can be used to try to normalise other shaped distributions. If your data are negatively skewed, squaring each of the values can make them form a more normal distribution. For a description of other transformations that can correct different shaped distributions, see Stevens (1996: 238).

4.7 Multivariate distributions

In the first half of this chapter, we were only interested in univariate distribution. We will now use the principles illustrated there to illustrate aspects of the more complex *multivariate* distributions. Although the distributions become a little more complicated, the principles remain the same.

A multivariate distribution is a distribution that contains more than one variable, and a multivariate distribution requires that we consider what are known as the *joint distributions*. You will recall from Chapter 1 that residuals are the difference between the score predicted by the model and the actual scores that we have for the variables (they are the extent to which the model is wrong). If we consider the mean to be a one-parameter model, then the residuals are the differences between the mean and each value.

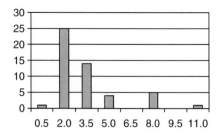
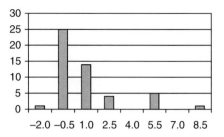

FIGURE 4.13   FIGURE 4.14

The residual for each value is the amount the value differs from the mean score *if we only used the mean as our prediction of what that score would be*. At the start of Chapter 1, we only had the mean to use as a predictor of our heights, but you managed to make a good guess based on that limited information.

Figure 4.13 shows a histogram of the frequency distribution of the data that were presented in Table 4.14. Figure 4.14 shows a histogram of the residuals, the differences between the mean score and each value.

You can see from these two figures that the distributions have the same shape. In fact the graphs are identical, with the one difference being that the graph of residuals has been shifted to the left. This is an important thing to note, because it shows that the distribution of residuals of a single variable is the same as the distribution of the original variables and so everything that we said about distributions in the preceding section applies to distributions of residuals in this section. It follows that we can talk about multivariate distributions in much the same way, because we can consider them in terms of the distributions of their residuals. And this, as we will see, is a lot easier.

In the previous section, we looked at the mean, which you will recall is a least squares estimator. The slope coefficients that we are dealing with are also least squares estimators, and so much of what we said about the simple case of the mean is true of the more complex case of the slope.

If we have a bivariate regression with one dependent variable and one independent variable, we are no longer (very) interested in outliers in the univariate distributions – we are now interested in outliers in the joint distribution. We are also interested in the shape of the joint distribution.

The assumptions that are made are listed below:

1. At each value of the dependent variable, the distribution of the residuals is normal.
2. The variance of the residuals at every set of values for the independent variable is equal. This assumption that the variance is equal is called homoscedasticity (and if it should happen that the variance is not equal, that our assumption of equality is not satisfied, then that condition is called heteroscedasticity).
3. At every possible value of the dependent variables, the expected (mean) value of the residuals is equal to zero. In the bivariate case, this assumption

86 APPLYING REGRESSION AND CORRELATION

means that the relationship between the independent variable and the dependent variable should be linear.
4. For any two cases, the expected correlation between the residuals should be equal to zero. This is referred to as the independence assumption, or a lack of autocorrelation.

4.7.1 Assumption 1

At each value of the dependent variable, the distribution of residuals is normal.

The first way to check this assumption is to examine whether the distribution of residuals is approximately normal. If the distribution differs from normality, this assumption is violated. Remember that this non-normality can occur because of outliers, or because of skew and/or kurtosis.

4.7.1.1 Outliers

A hypothetical study has been carried out examining the IQ of thirty couples. The purpose of the study was to test the hypothesis that people seek partners of similar intelligence to themselves. The data are presented in Table 4.15.

If we were to draw histograms of the univariate distributions (Figure 4.15), we would find that they do not appear to have any outliers, and seem to follow approximately a normal distribution.

However, when we plot a scattergraph, it becomes obvious that there is an outlier: the scattergraph is shown in Figure 4.16 with the outlier labelled as couple number 11. We should now investigate to find out if this outlier occurred as the result of a recording error, a data entry error, or if it is a real result that is not accounted for by our theory, just as we did with the univariate case when we examined the mean.

The technique of drawing a scattergraph and visually examining it works fine with two variables, but when there are more than two variables, the technique is no longer satisfactory. If we are using paper, it becomes difficult to draw a graph in more than two dimensions (and virtually impossible using any technique in more than three) to examine whether there are any outliers, so we need a different technique. We need a numerical measure that we can use to see if a case is an outlier, in much the same way as we used the standardised residual when we were looking at the mean. To test for outliers, we calculate a regression equation, and see how far the score for each couple is from their predicted score.

Table 4.16 shows the output from a regression calculation, where male IQ is the dependent variable and female IQ is the independent variable. This shows that if we know the female's IQ the best estimate of a male IQ in the couple is given by:

$$\text{Male IQ} = 0.622 \times \text{female IQ} + 35.2$$

TABLE 4.15 Dataset 4.2a

Couple	Male IQ	Female IQ
1	74	75
2	126	125
3	134	141
4	96	77
5	78	91
6	66	64
7	80	66
8	58	55
9	92	87
10	92	95
11	60	138
12	98	111
13	90	89
14	118	134
15	86	109
16	68	86
17	118	122
18	98	116
19	114	107
20	116	111
21	84	71
22	92	80
23	100	100
24	144	136
25	86	71
26	128	104
27	86	77
28	110	114
29	112	118
30	78	68

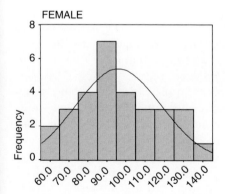
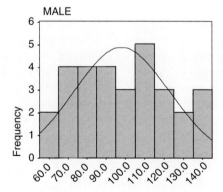

FIGURE 4.15

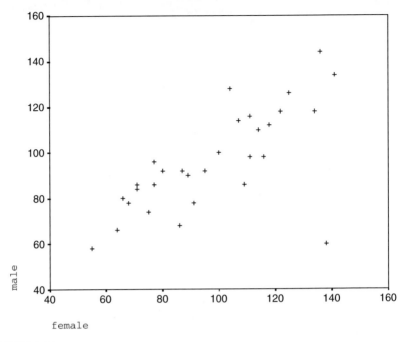

FIGURE 4.16

TABLE 4.16

	Slope (*b*)	Std error of slope	Standardised slope (beta)	*t*	Sig.
Constant	35.200	12.360		2.848	0.008
female	0.622	0.123	0.692	5.073	0.000

We can use this to calculate a predicted value for each couple, and then we can use this to calculate a residual, in the same way as we could have used the mean to calculate a residual. To do this, we use the female partner's IQ to predict a value for the male partner's IQ. We then calculate the difference between the predicted value for each male and the actual value that the male scored for his IQ. Figure 4.17 shows in graphical form how these figures are calculated. Each cross represents a couple. The regression line is the line of best fit and the perpendicular distance between each point and the line shows the difference between the predicted value and the actual value – the residual.

Table 4.17 shows the original male and female IQ for each couple, along with the predicted male IQ (based on the model), and the residuals.

Given the small size of this dataset it is easy to see that the residual for couple 11 is unusually high (and, of course, it helped that we knew where to look). In a larger dataset, we might want to see a histogram of residuals to easily identify whether there are any outliers (and most computer packages will draw this automatically). The histogram of the residuals, which shows that there is an outlier, is shown in Figure 4.18.

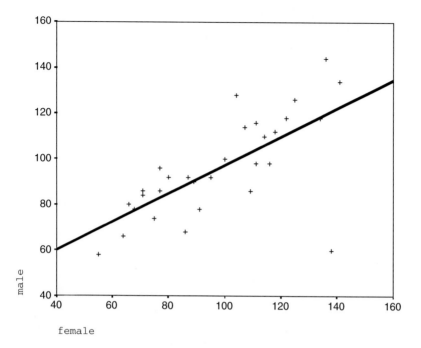

FIGURE 4.17 *Scatterplot showing line of best fit*

Another way of graphically thinking about residuals is to think of the line of best fit as being a straight, horizontal line, and the residuals as being a certain distance above or below that line. We can draw a scatterplot with the predicted value of the variable on the *x*-axis (the horizontal axis) and the residual on the *y*-axis (the vertical axis). Many packages will produce this graph automatically, and an example is shown in Figure 4.19.

This graph appears to be very similar to the scattergraph shown in Figure 4.17, but it has one very large advantage that we will deal with now (and a couple of other advantages that we will look at in a few pages). Whereas the scattergraph of Figure 4.17 was limited to two variables – one dependent variable and one independent variable – a residual plot such as this one uses the residuals, not the original variables. Because the plot uses the residuals, it is possible to use it to check for outliers where we have more than one independent variable.

Table 4.18 contains a (fictional) dataset where 40 people have been assessed using four measures, a life events scale (called `events`), a hassles (`hass`) scale, a social support scale (`supp`) and a depression scale (`dep`). We are interested in seeing the extent to which the three variables of life events, hassles and social support are able to predict current levels of depression.

If the frequency distributions are analysed, it can be seen that the distributions are all fairly normal, with no outliers. The bivariate scatterplots, shown in Figure 4.20, also reveal that there are no bivariate outliers.

TABLE 4.17 Dataset 4.2b

Couple	Male IQ	Female IQ	Predicted male IQ	Residual
1	74	75	81.81	−7.81
2	126	125	112.89	13.11
3	134	141	122.83	11.17
4	96	77	83.06	12.94
5	78	91	91.76	−13.76
6	66	64	74.98	−8.98
7	80	66	76.22	3.78
8	58	55	69.38	−11.38
9	92	87	89.27	2.73
10	92	95	94.24	−2.24
11	60	138	120.97	−60.97
12	98	111	104.19	−6.19
13	90	89	90.51	−0.51
14	118	134	118.48	−0.48
15	86	109	102.94	−16.94
16	68	86	88.65	−20.65
17	118	122	111.02	6.98
18	98	116	107.30	−9.30
19	114	107	101.70	12.30
20	116	111	104.19	11.81
21	84	71	79.33	4.67
22	92	80	84.92	7.08
23	100	100	97.35	2.65
24	144	136	119.73	24.27
25	86	71	79.33	6.67
26	128	104	99.84	28.16
27	86	77	83.06	2.94
28	110	114	106.05	3.95
29	112	118	108.54	3.46
30	78	68	77.46	0.54

The results of the regression analysis are shown in Table 4.19. These results show that only life events are having a significant effect on levels of depression.

However, if we carry out a regression analysis and plot the scattergraph of the predicted values against the residuals, as shown in Figure 4.21, then we find that there is an outlier. This outlier is case 3 in the dataset. If you actually have a look at the values for the variables for case 3, you can see that the life events score of 150 ($z = 0.6$) is fairly moderate. The hassles score is again fairly high ($z = 1.7$), but not enormously high, and not enough to call it an outlier. The social support score is low ($z = -1.5$), but not so low as to be unusual, and the depression score is low ($z = -1.8$), but also not so low as to be unusual. What *is* unusual is the combination of all those scores: when we consider all of the scores together, case 3 becomes an outlier.

4.7.1.2 Types of residuals

When we were discussing the mean, we saw that residuals could be raw or standardised (adjusted so that SD = 1) and that the residuals could be calculated from a dataset in which a particular individual case was deleted.

ASSUMPTIONS IN REGRESSION ANALYSIS 91

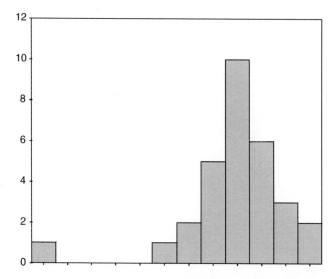

FIGURE 4.18

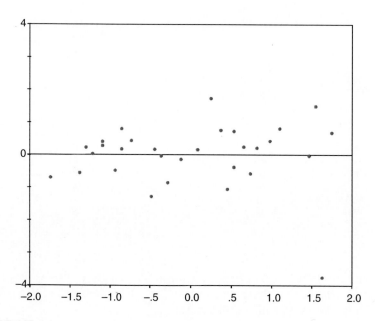

FIGURE 4.19

TABLE 4.18 Dataset 4.3

Case	events	hass	supp	dep
1	120	46	10	150
2	185	48	33	160
3	150	60	10	70
4	110	55	23	150
5	94	41	47	90
6	156	54	40	130
7	55	31	48	120
8	132	33	41	120
9	141	31	32	150
10	157	59	38	160
11	202	53	13	180
12	107	51	39	120
13	148	39	37	140
14	180	43	13	120
15	88	43	51	100
16	108	43	30	110
17	111	40	5	120
18	122	46	64	80
19	162	55	60	140
20	150	51	28	120
21	148	45	47	90
22	103	50	18	130
23	155	57	19	140
24	134	46	45	130
25	116	36	46	90
26	159	60	18	120
27	122	46	32	90
28	149	36	39	90
29	86	42	39	80
30	100	36	44	110
31	127	50	15	150
32	130	54	49	100
33	95	56	33	90
34	52	33	49	110
35	91	45	41	110
36	148	43	20	110
37	150	39	21	90
38	143	45	24	160
39	129	37	39	140
40	179	55	10	150

Residuals can be treated in the same ways, in the case where we have more than one variable and are trying to detect multivariate outliers. We will now consider the types of outliers that exist.

4.7.1.2.1 Unstandardised residuals (RESID)
Unstandardised residuals are the 'raw' residuals that we encountered earlier. They are simply the difference between the predicted value and the actual value for the particular cases.

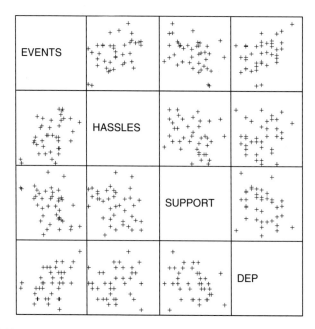

FIGURE 4.20

TABLE 4.19

	Slope (b)	Std error of slope	Standardised slope (beta)	t	Sig.
Constant	97.77	29.980		3.261	0.002
events	0.29	0.138	0.355	2.081	0.045
hassles	−0.05	0.536	−0.015	−0.093	0.927
support	−0.380	0.292	−0.210	−1.298	0.202

4.7.1.2.2 Standardised residuals (ZRESID)

If we do not know the underlying scale of a variable, it is difficult to interpret the residual (we saw this with the mean). The solution using the mean was to standardise the scores to give them a mean of 0 and standard deviation of 1. We encountered the same problem with the residuals in a regression equation, and use the same solution. If we standardise the residuals to give them a mean of 0 and a standard deviation of 1, we know the scale of measurement and therefore we are able to interpret the residual.

4.7.1.2.3 Studentised residuals (SRESID)

The first two types of residuals that we discussed assume that the residuals have the same variance at every value of the predicted variable. To avoid this assumption, the studentised residual makes a correction based on the estimated variance of the residual at that value of the predicted variable.

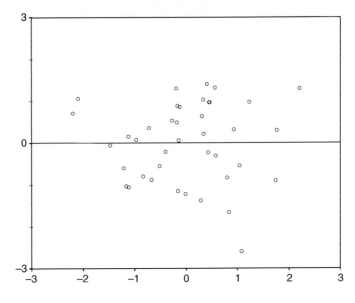

FIGURE 4.21

4.7.1.2.4 Deleted residuals (DRESID)

When calculating residuals we are interested in the difference between the predicted value and the actual value. When we calculate the predicted value, we include the (potential) outlier. This means that the outlier has an influence on the predicted value, making it seem less extreme than if we calculated the estimate without including the value of the outlier. Because outliers may have undue influence, deleted residuals and studentised deleted residuals are calculated for each case, based on the predicted value if that case were excluded from the analysis.

4.7.1.2.5 Comparison of residuals

Table 4.20 shows each of the different types of residual for case 3 in the dataset. The raw residual seems large, but this is difficult to interpret because it is affected by the measurement scale of the dependent variable – it seems large, but we just do not know unless we know the scale. The deleted residual (DRESID) is larger than the raw residual, which indicates that the outlier is influencing the position of the line – if the outlier were having no effect, the deleted residual and the raw residual would be equal. The standardised and studentised residuals make the size of the residual much clearer because we now know that the residuals have a mean of 0 and a standard deviation of 1. The absolute values of 2.6 (ZRESID) and 2.8 (SRESID) are both quite high. The studentised deleted residual is higher than the studentised residual, again showing that the outlier is having an effect on the position of the regression line.

TABLE 4.20 *Comparison of residuals*

Residual	Value
Raw (RESID)	−63.9
Deleted (DRESID)	−73.9
Standardised (ZRESID)	−2.6
Studentised (SRESID)	−2.8
Studentised deleted (SDRESID)	−3.1

4.7.1.3 Interpreting outliers

Outliers do not always have an influence on the results of our regression analysis. Cases that do have an effect on the outcome of the calculation are referred to as *influential cases*. In the examples given earlier in the chapter, all the outliers affected our final model, but if we were to look at the more complex bivariate and multivariate analysis found in regression, we would see that not all outliers are influential. However, we will examine influence statistics only in the bivariate case in which there is one independent variable. Everything that we say about the relatively simple bivariate case generalises to the multivariate case, but the complexity of the calculations increases exponentially.

We will consider two general types of outlier detection statistics: they are *distance statistics* and *influence statistics*. We will look at three types of distance statistics, leverage, Mahalanobis distance and Cook's D, and four influence statistics, DfBeta, standardised DfBeta, DfFit and standardised DfFit.

4.7.1.4 Distance statistics

4.7.1.4.1 Leverage

Leverage is so called because a lever can be imagined as pulling on the regression line to make it go in a certain direction. Leverage is calculated using only the values of the independent variable(s), so it is possible that an influential case may not be detected solely by examining its leverage statistic.

The calculation for the leverage statistic (called h) for the ith case is shown below. It can be seen that any two cases that share the same value for X (the independent variable) will have the same leverage value. Leverage values have a maximum possible value of 1.

$$h_i = \frac{1}{N} + \frac{(x_i - \bar{x})}{\sum x^2}$$

Hoaglin and Welsch (1978) recommend the rule of thumb whereby a cut-off value for leverage is given by:

$$\frac{2(k+1)}{N}$$

where k is the number of independent variables, and N is the number of participants.

4.7.1.4.2 Mahalanobis distance

The Mahalanobis distance is calculated from the leverage value. The advantage of the Mahalanobis distance is that it is possible to use the distances as a value with a known distribution, which can then be tested for significance, by finding its associated probability (although not all statistics packages do this automatically). If there is only one independent variable, the Mahalanobis distance is the square of the standardised value of the variable. For cases with multiple independent variables, it is calculated using:

$$MD_i = (N-1)\left(h_i - \frac{1}{N}\right)$$

4.7.1.4.3 Cook's D

Whereas the leverage value and the Mahalanobis distance used only the values of the independent variables in their calculation, Cook's D uses both the independent variable and the dependent variable. Cook's D uses both the value of the studentised residual and that of the leverage statistic to calculate a distance. The equation is:

$$D_i = \left(\frac{\text{SRESID}_i}{k+1}\right)\left(\frac{h_i}{1-h_i}\right)$$

Because Cook's D is a product of the leverage and the residual, if either one of leverage or SRESID is low, then D will be low. If both values are high, the value for D will be high.

4.7.1.5.1 Comparison of distance statistics

The comparison is given in Table 4.21.

TABLE 4.21 Comparison of distance statistics for case 3

Statistic	Value
Leverage	0.11
Mahalanobis distance	4.3
Cook's D	0.30

The summary statistics for all cases are given in Table 4.22. The leverage value and the Mahalanobis distance for case 3 are not the highest (nine cases have higher scores on these statistics). However, recall that these two statistics do not take into account the score for that respondent on the dependent variable, but only consider the independent variables. Cook's D considers both the dependent and the independent variable, if the value of Cook's D for case 3 has the highest score.

TABLE 4.22 *Summary of distance statistics*

	Minimum	Maximum	Mean	Std deviation
Leverage	0.002	0.207	0.075	0.046
Mahal. distance	0.092	8.075	2.925	1.780
Cook's D	0.000	0.303	0.032	0.051

4.7.1.7 Influence statistics

When we looked at the effects of outliers on the mean, we were not solely interested in whether a particular case was an outlier; we were also interested to find out what effect that outlier had on our model. We saw this with the mean, in Section 4.5.2.2. If one case can dramatically alter the model that we end up with, then that case is said to be having an undue influence.

We will look at two types of influence statistics, DfBeta and DfFit, both of which come in a standardised and an unstandardised form.

4.7.1.7.1 DfBeta and standardised DfBeta

DfBeta for a case is the difference between the value of beta (i.e. the slope) when the case is included, and the value of beta when the case is excluded (although see Pedhazur (1997: 52) for some elaboration on this). At this point, it should not surprise you to find that there is an unstandardised form, which can be more difficult to interpret, and a standardised form, which is easier to interpret. Because every case can have an influence on every parameter estimated in the model, it turns out that there are a very large number of values for DfBeta — one per case, per parameter estimate, in both the standardised and unstandardised forms. In the model that we have been examining, we estimate three slope coefficients and one intercept. There will be one value of DfBeta for each of these estimates, and one value of standardised DfBeta for each. These are referred to as dfbeta$_{ij}$ where the subscript i refers to the case number and j refers to the parameter estimated. The DfBeta or the intercept is called DfBeta$_0$ (the intercept being b_0).

Table 4.23 shows the values for the four DfBetas in the depression data, in both standardised and unstandardised forms, for the first 10 cases in the depression dataset that we have been considering. You can see that case 3 does not have very much influence on the intercept of the first slope coefficient (events), but does seem to be having a much larger effect on both of the other parameters (the slope coefficient for hass and supp).

A common recommendation is that a cut-off of:

$$\text{DfBeta} > \left| \frac{2}{\sqrt{N}} \right|$$

should be used to determine whether a case is influential. Given that we have a sample where $N = 40$, our cut-off should be:

TABLE 4.23 Values of DfBeta

Case	Unstandardised				Standardised			
	DfBeta$_0$	DfBeta$_1$	DfBeta$_2$	DfBeta$_3$	DfBeta$_0$	DfBeta$_1$	DfBeta$_2$	DfBeta$_3$
1	6.90	−0.02	−0.01	−0.09	0.23	−0.15	−0.02	−0.31
2	−4.50	0.04	−0.04	0.03	−0.15	0.32	−0.07	0.11
3	**5.16**	**0.03**	**−0.36**	**0.16**	**0.19**	**0.26**	**−0.74**	**0.62**
4	0.28	−0.04	0.17	−0.05	0.01	−0.32	0.32	−0.17
5	−1.21	0.01	0.00	−0.02	−0.04	0.07	0.00	−0.06
6	−1.50	0.00	0.02	0.01	−0.05	0.03	0.03	0.04
7	11.10	−0.04	−0.10	0.00	0.37	−0.32	−0.19	0.00
8	0.31	0.00	−0.01	0.00	0.01	0.01	−0.02	0.00
9	7.81	0.03	−0.23	−0.02	0.26	0.21	−0.42	−0.05
10	−12.80	0.02	0.21	0.07	−0.43	0.11	0.39	0.25

$$\text{DfBeta} > \left|\frac{2}{\sqrt{40}}\right| = \frac{2}{6.32} = 0.3165$$

4.7.1.7.2 DfFit and standardised DfFit

DfFit is very similar to DfBeta, but rather than looking at the change in the parameter estimates that occur as a result of excluding a case, it examines the change in the predicted value of a case, when that case is excluded — that is, the change in the 'fit' of the model. To calculate DfFit the model is estimated using all of the cases, and then estimated again, with the first case excluded. This process is repeated for every case DfFit is the difference between the predicted value for a case when the case is included in the model, and the predicted value that would be calculated for that case when the case is removed from the model.

If we estimate the regression equation for the depression data, we find that the predicted values are calculated as:

dep = (0.28 × events) + (0.05 × hass) + (−0.38 × supp) + 97.7

The predicted value for case 3 is therefore:

dep = (0.28 × 150) + (0.05 × 60) + (−0.38 × 10) + 97.7 = 133.9

We can also estimate the regression equation again, excluding case 3. This time we find the parameter estimates given in Table 4.24.

We can recalculate the predicted value for case 3 with this new model, as being:

dep = (0.26 × 150) + (0.31 × 60) + (−0.54 × 10) + 92.6 = 144.8

The predicted value for case 3 has altered from 133.9 to 144.8, a fairly large shift, although again this is hard to interpret without knowledge about the

TABLE 4.24 Slope coefficients with case 3 excluded

	Slope (b)	Std error of slope	Standardised slope (beta)	t	Sig.
Constant	92.609	26.973		3.433	0.002
events	0.255	0.124	0.330	2.054	0.048
hass	0.309	0.495	0.096	0.624	0.537
supp	−0.541	0.268	−0.305	−2.023	0.051

dispersion of the dependent variable, so to make this easier to interpret, standardised DfFit is used.

Table 4.25 shows the values of DfFit and standardised DfFit for the first 10 cases of the depression dataset. You can see that removing a case, other than case 3, has a very small effect on the predicted values.

TABLE 4.25

Case	DfFit	Standardised DfFit
1	2.94	0.36
2	3.06	0.37
3	−9.97	−1.23
4	3.81	0.48
5	−1.01	−0.16
6	0.47	0.07
7	5.62	0.55
8	0.19	0.02
9	4.42	0.47
10	4.49	0.55

The difference in fit is calculated as the change in the predicted value that occurs when the case is excluded. The rule of thumb that should be used is to examine cases that have a cut-off of:

$$\text{DfFit} > \left| \frac{2}{\sqrt{k/N}} \right|$$

4.7.2 Assumption 2

The variance of the residuals at every set of values for the independent variable is equal.[7] *This assumption is referred to as homoscedasticity. The condition of violating this assumption is called heteroscedasticity.*

Heteroscedasticity is spotted using the same type of residual scatterplots that we examined when looking for outliers. In that case, we were looking for specific points that were lying outside the general distribution; in this case we are looking at the general pattern formed by the residuals.

Consider Figure 4.22, which shows a scatterplot of the residuals in a regression calculation on the y-axis and the predicted values on the x-axis, the

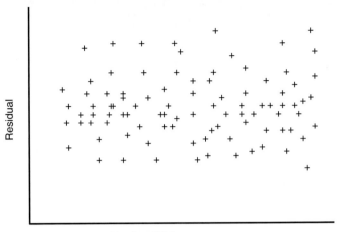

FIGURE 4.22

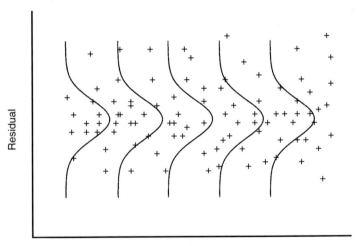

FIGURE 4.23

same type of plot that we used to examine the possibility of multivariate outliers.

We can see that there are no outliers in this dataset, but we are now interested in the variance of the residuals at each level of the predicted values. We are interested in whether the variances of the residuals, at each predicted value of the dependent variable, are equal. Figure 4.23 shows the plot again with a series of normal distributions placed upon it. If we were to draw a histogram of the residuals taken from any point along the x-axis, we would

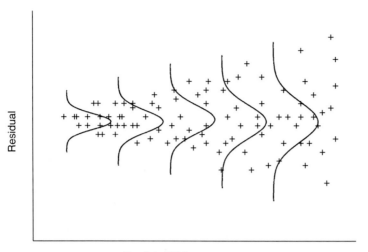

FIGURE 4.24

find that the variances of the distributions would be approximately equal, as shown in the figure.

Figure 4.24 shows a scatterplot of residuals in which the assumption of homoscedasticity has been violated. Again, histograms showing normal distributions have been drawn at various points along the graph. You can see that the variances of these distributions vary along the length of the x-axis.

4.7.2.1 The implications and meaning of heteroscedasticity

From a purely statistical viewpoint, violation of the heteroscedasticity assumption is not as serious as violation of certain other assumptions. The parameter estimates of a heteroscedastic dataset will not be altered, in much the same way that the mean is not altered if a distribution is not skewed, but is positively or negatively kurtosed (as we saw in the previous section). Instead, as with the mean again, the standard errors of the estimate will be inaccurate, and therefore any calculations of significance will be wrong to some extent.

The meaning of heteroscedasticity is more important from the perspective of the specification of the model, because the presence of heteroscedasticity means that the model has been mis-specified. Table 4.26 shows a dataset in which people who worked for a company were able to donate, from their salary, money to the adopted charity for the year.

Three measures were taken: first, the amount of money that the person earned in that year, labelled cash in the table; second, the level of importance the individual assigned to the charity, labelled as import; and finally, the amount given, called given.

If we calculate a regression equation, using given as the dependent variable and cash and import as the independent variables, we find $R^2 = 0.599$, $p < 0.0005$.

TABLE 4.26 Dataset 4.4

Case	cash	import	given
1	16	9	43
2	15	12	56
3	17	11	51
4	14	8	50
5	11	11	25
6	20	14	69
7	15	13	46
8	19	10	56
9	21	12	66
10	7	12	20
11	18	8	41
12	16	6	33
13	20	9	58
14	17	8	38
15	14	12	38
16	18	10	50
17	20	11	64
18	12	9	36
19	17	9	55
20	17	13	62
21	11	5	50
22	17	11	53
23	10	9	45
24	20	8	58
25	19	13	64
26	15	11	42
27	13	7	40
28	9	10	25
29	21	10	62
30	10	10	49
31	19	10	57
32	10	9	44
33	16	11	49
34	15	7	37
35	11	8	38
36	18	12	61
37	16	15	62
38	17	8	51
39	9	9	34
40	16	10	44

The parameter estimates are given in Table 4.27.

It can be seen that we can make a very good prediction of how much money someone will give to a charity if we know how much they earn and the importance they attach to the work done by that charity.

When we examine the residual plot shown in Figure 4.25 (as, of course, we always should), we find something rather curious. The data have violated the assumption of homoscedasticity. At the lower end of the predicted values, the variance of the residuals is relatively high, but at the higher end of the predicted values, the variance is much lower. We have already said that from

TABLE 4.27

	Slope (*b*)	Std error of slope	Standardised slope (beta)	*t*	Sig.
Constant	0.239	6.833		0.035	0.972
cash	2.283	0.329	0.712	6.943	0.000
given	1.265	0.565	0.229	2.238	0.031

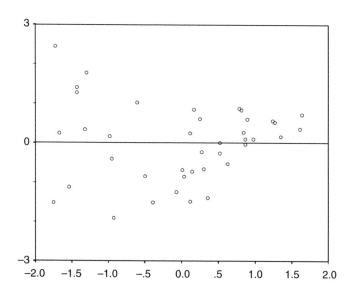

FIGURE 4.25

a statistical point of view this does not concern us a great deal, but from a theoretical standpoint we should be concerned.

The presence of heteroscedasticity, such as is shown in Figure 4.25, means that there is a more complex relationship between the variables than we have modelled. The fact that an individual earns a larger amount of money is not, on its own, sufficient for us to predict that that individual will donate more money to the charity. Similarly, that the individual considers the work that the charity does to be important is not sufficient, on its own, for us to predict that that individual will donate more money to the charity. An individual must earn more money *and* consider the work the charity does to be important before that individual will donate more money to the charity. The effect of all this is that the extent to which earning money is important as a predictor of donation depends upon the level of importance attached to the charity by the individual. Our model therefore missed out a very important theoretical predictor of donation – the interaction effect between different variables. G.H. McClelland (personal communication) refers to this effect as being 'different slopes for different folks', and we will explore it in some depth in Chapter 6. For now, to demonstrate the point we will present the scatterplot of importance and generosity, split between high and low earners, in

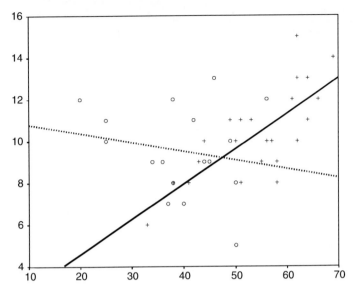

FIGURE 4.26

Figure 4.26; importance is shown on the *x*-axis and amount of money on the *y*-axis. The crosses in the figure represent the higher earners, the solid line shows the line of best fit for the high earners, and the circles represent the low earners – the dotted line shows the line of best fit for the lower earners. You can see that amongst the higher earners, the more importance people assign to the charity the more money they give – the higher earners are clustered quite tightly around the line of best fit, indicating a strong, positive relationship. The low earners do not show any such relationship – the line of best fit slopes downwards, slightly, and they are not clustered tightly around the line (in fact the slope is not significant). We have, as McClelland put it, different slopes for different folks.

We will consider how to analyse this type of relationship in further detail in Chapter 7.

4.7.3 Assumption 3

At every possible value of the dependent variables, the expected (mean) value of the residuals is equal to zero. In the bivariate case, this means that the relationship between the independent variable and the dependent variable should be linear.

The expected value for a residual is the value that you would expect that residual to have *if you had no other information* – think back to the start of the book (or read it if you jumped straight in here) to determine why this is the case.

4.7.3.1 Example with one independent variable

Take, for example, the dataset shown in Table 4.28, which has two variables.

TABLE 4.28 Dataset 4.5

x	y
1	0
2	30
3	48
4	60
5	70
6	78
7	85
8	90
9	95
10	100
11	104
12	108
13	111
14	115
15	118
16	120
17	123
18	126
19	128
20	130
21	132
22	134
23	136
24	138
25	140
26	141
27	143
28	145
29	146
30	148
31	149
32	150

If a regression analysis is carried out on these data using x as the independent variable and y as the dependent variable, we find that $R^2 = 0.844$. Thus, the independent variable accounts for a great deal of the variance in the dependent variable. This means that we can predict the dependent variable with a high degree of accuracy, if we know the value for the independent variable. If we draw a scatterplot of the two variables (Figure 4.27), the picture becomes clearer.

The graph shows that there is a consistent relationship between the independent variable and the dependent variable, but that it is not linear. The line of points is very smooth, and therefore by using the graph we can make an almost perfect prediction of a person's score on the dependent variable given their score on the independent variable. Because we can make a very accurate (almost precise) prediction, the value of R^2 *should* be very close to 1.00.

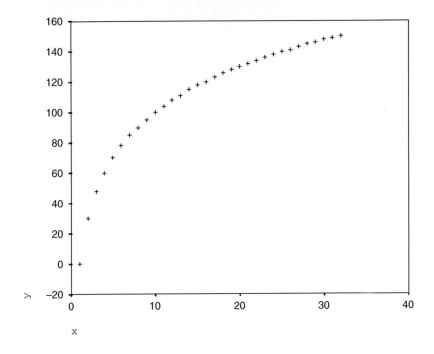

FIGURE 4.27

As this value is close to 1.00, this means that we have found that we can make accurate predictions, but not perfect predictions. By using a linear relationship we have lost some of our ability to predict the dependent variable. Instead of using a linear relationship, we could use a non-linear relationship.

In the regression analysis that we have so far carried out, we have used the formula:

$$y = bx + c$$

The dependent variable (y) is a linear function of the dependent variable (x). The predicted increase in y when x increases by one unit remains constant across the whole range of x. In all of the previous examples that we have examined, this has been the case: in Chapters 1 and 2, reading one textbook had a constant effect on a person's marks, whether it was the first book they read or the fourth book they read. The trouble with this type of assumption is that it will often turn out not to be the case. This assumption can be violated for two reasons.

First, there is the *ceiling effect*: reading the 20th book will not have the same effect as reading the first. When you have full marks, you have reached the *ceiling*, and it is not possible to get any more marks. Statistically, this is not such a major problem as long as no one does read 20 books, and as long as you do not try to make predictions about what would happen if someone did read

20 books. (Attempting to make predictions beyond a reasonable range is called *extrapolation*, and can lead to this type of error.)

Second, violation of this assumption can occur when the effect is fundamentally non-linear – that is, it has a different effect at different values. The value people assign to cash is an example of a non-linear effect. How much you want some money depends, at least in part, on how much money you have already. When we (and possibly you) were young, a very small amount of money from our parents could tempt us to do all sorts of tasks – we would happily drop whatever we were doing and make a cup of tea, if we were offered some meagre sum, such as 50 pence.[8] Now that we are older and have more money, being given 50 pence would not tempt us to do very much. But when you are older and have more money, you would very much like to be given £1000 and would probably do quite a lot to gain it: you would, for example, go to work for eight hours a day, five days a week, for a number of weeks (we do, and so do a lot of people we know). Most people who find themselves in receipt of large amounts of money (say, by winning the lottery) suddenly find that £1000 is not enough to go to work any more to get it.[9]

Table 4.29 shows a dataset containing details of a sample of houses in a large city. The table shows the size of the floor area of the house (in square metres, called `area`) and the price that was paid for those houses (in thousands of pounds, called `price`). When we try to predict the value of house prices using size, with a regression equation, we find the results shown in Table 4.30.

The expected value of the dependent variable is therefore given by:

$$\text{price} = \text{size} \times 3.9 + (-124.8)$$

This would suggest that adding 1 square metre of floor space would add approximately £4000 to the price. If we examine the scatterplot in Figure 4.28, which shows the predicted values plotted against the residuals, we can see that there is something curious going on. The data appear to show homoscedasticity; that is, the variance is similar across the range of the dependent variable. The expected value of the residuals (i.e. the mean value of the residuals) at every point along the range of the dependent variable is not equal to zero. At the lower and the higher end the values of the residuals are above zero and on the central range the values of the residuals are below zero.

This has occurred because there is a *non-linear* relationship between the two variables. This can be further explored using a scattergraph of the two variables (Figure 4.29) – to which we can add the line of best fit. It appears from this graph that the relationship between the two variables does not increase at a constant rate. When house sizes are smaller, adding floor space makes a small difference to the price of the house. When houses are larger, adding floor space makes a larger difference to the price of the house.

This is shown in Figure 4.30, where two lines of best fit have been drawn – the crosses show the houses of less than 100 square metres, and the dashed

APPLYING REGRESSION AND CORRELATION

TABLE 4.29

Case	area	price
1	72	156
2	98	153
3	92	230
4	90	152
5	44	42
6	46	157
7	90	113
8	150	573
9	94	202
10	90	261
11	90	175
12	66	212
13	142	486
14	74	109
15	86	220
16	46	186
17	54	133
18	130	360
19	122	283
20	118	380
21	100	185
22	74	186
23	146	459
24	92	167
25	100	171
26	140	547
27	94	170
28	90	286
29	120	293
30	70	109
31	100	205
32	132	514
33	58	175
34	92	249
35	76	234
36	90	242
37	66	177
38	134	399
39	140	511
40	64	107

TABLE 4.30

	Slope (*b*)	Std error of slope	Standardised slope (beta)	*t*	Sig.
Constant	−124.723		−3.3	0.002	0.034
size	3.966	0.862	10.5	0.000	<0.001

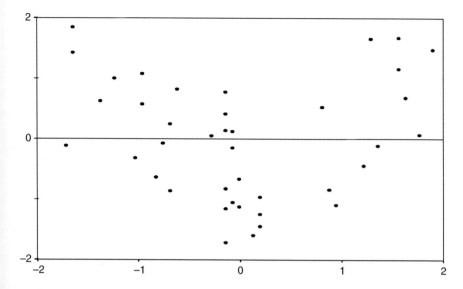

FIGURE 4.28

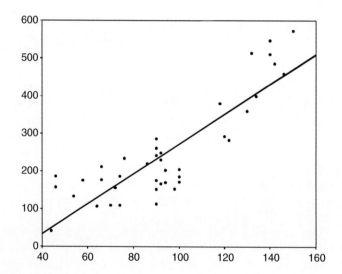

FIGURE 4.29

line shows the line of best fit for these houses. The solid circles show the houses with floor space greater than 100 square metres, and the solid line shows the line of best fit for these houses.

In Chapter 6, we will look at the approaches that can be used to analyse non-linear relationships, but for now we will say that we *usually* do not have to worry about non-linearity to a great extent.

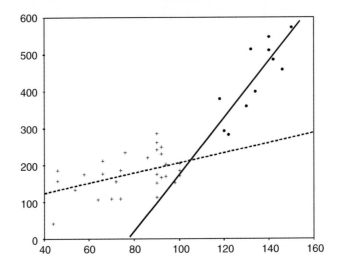

FIGURE 4.30

4.7.4 Assumption 4

For any two cases, the expected correlation between the residuals should be equal to zero. This is referred to as the independence assumption, or a lack of autocorrelation.

This is the most difficult assumption that we will deal with in regression analysis, and the one which tends to be ignored more than the others – although this is because it can be difficult to detect whether this assumption has been violated. In Chapter 8 we will expand on the independence assumption, and look at the approaches that can be used when this assumption has been violated, so this section will remain short.

Autocorrelation occurs when a variable correlates with itself: if the cases are autocorrelated, then they are related to one another – they are not independent.

There are two occasions when this can occur. The first is in a time-series design, where multiple measures of the same entity are assessed; this situation is usually referred to as autocorrelation. The second occurs when the units of analysis can be grouped or clustered in some way, often by geographical area; this situation is usually referred to as non-independence.

If violation of this assumption is suspected, it can be difficult to determine whether it has actually occurred. In this section we will focus on avoiding situations where this may arise.

4.7.4.1 Time-series designs

In a time-series design, the sample that is used is a repeated measurement of an individual over time. For example, the independent variable may be the

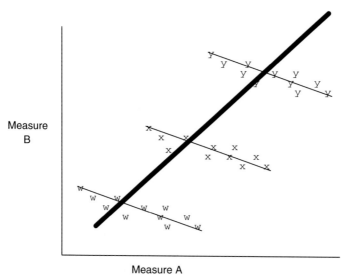

FIGURE 4.31

amount of food that a person ate every week for a period of one year, and the dependent variable may be the weight of that individual. Because an individual's weight at week 1 and their weight at week 2 are more related than their weight at week 1 and their weight at week 50, this is likely to violate the autocorrelation assumption.

4.7.4.2 Clustered sampling designs

A clustered sampling problem is a more common type of problem than a time-series design in social science research. This occurs when people are related in some way (usually geographically), and are therefore not independent of one another. The area in which this has had the largest impact is school effectiveness studies (Aitkin & Longford, 1986): in this type of study, the participants are clustered within classes, and are therefore not independent of one another. Additionally, classes are clustered within schools, schools may be clustered within districts, and they in turn may be clustered within cities. Figure 4.31 (reprinted from Dunn & Miles, 1996) shows how this effect may distort results in regression analysis. The figure shows a scattergraph of the relationship between two measures (referred to here as Measure A and Measure B). Superficially, it appears that the line of best fit would be somewhere in the region of the thick black line, indicating a positive relationship.

However, the sample was made up of three separate clusters – these may have been schools, countries or occupational groups, which are labelled w, x and y in the figure. You can see that if we consider the clusters individually, the correlation within each cluster is actually negative; by failing to consider the structure of the data, we have found that the correlation has been reversed. These types of data are referred to as being hierarchical (e.g. by Bryk & Raudenbush, 1992) or multilevel data (e.g. by Goldstein, 1995).

Notes

1 Haggis, according to the *Monty Python Big Red Book* (Chapman et al., 1991), 'is a kind of stuffed black pudding eaten by Scots and considered by them not only a delicacy but fit for human consumption. The minced heart, liver and lungs of a sheep, calf or other animal's inner organs are mixed with oatmeal, sealed and boiled in maw in the sheep's intestinal stomach-bag and . . . Excuse me a minute. Ed.'

2 And also the name of a drug designed to help people lose weight – Leptin.

3 For an account of the rivalry between these two pioneers of statistics, see Gigerenzer and Swijtink (1990).

4 At a conference recently one of us heard of a (probably apocryphal) statistician who, when asked where his data came from, replied 'I found them on my desk.' This example demonstrates why this is not an approach that should be advocated.

5 We are reminded of a Woody Allen joke about working for a stripper. When he gets to the front of the queue, he is asked how much he would like to pay to have the job.

6 For this type of study, a special type of regression analysis is used, called survival analysis. Discussion of this is beyond the scope of this book, but the interested reader is directed to Collett (1994).

7 If you have read Chapter 3 you might realise that this assumption is equivalent to the homogeneity of variance assumption that is made by ANOVA and *t*-tests.

8 This never happened. Our parents simply demanded tea. But it is just an example.

9 The non-linear value of money explains, at least in part, why people buy lottery tickets, when, according to statisticians, it is not a sensible thing to do. If the value of money were linear, it would not be a sensible thing to do, and people would probably do it less. In the same way that the value of £1000 is diminished to the lottery winner, the value of £1 is very much diminished to us to such an extent that it is very close to zero. The probability of winning the UK lottery is (approximately) 14 000 000 to 1, and winners usually win less than this sum (in the two weeks preceding the writing of this section, the winners won, on average, £1 030 000. The over-empirical statistician would say that this shows how foolish people are – the expected returns are less than the expected outlay on tickets. The psychologically informed statistician (or statistically informed psychologist) realises that this calculation is erroneous, as it assumes that the value of £1 is consistent across the whole range of possible sums of money. To most people, £1 is a fairly small, indeed trivial, sum of money, therefore the expected return is not £1 million, for an expected outlay of £14 million, rather the expected return is 'lots' for an expected outlay of 'nothing'.

Further reading

Fox (1991) *Regression diagnostics* and Berry (1993) *Understanding regression assumptions* are both short books which contain exactly what they say on the cover. Tukey (1977) *Exploratory data analysis* is an excellent and highly detailed text about different techniques for exploring data, checking for outliers and checking distributions.

5 Issues in regression analysis

This chapter will focus on some of the trickier issues in regression analysis. Some of the issues that we examine in this chapter would normally be listed under 'assumptions'. We distinguish between the assumptions and issues on the basis that the things we call assumptions are less contentious, and a little more cut and dried, than the things we call issues. These issues require more judgement on the part of the data analyst (i.e. you), and sometimes mean that there is no single satisfactory solution. We will be considering four issues in this chapter: causality, sample size, collinearity and measurement error.

5.1 Causality

Causality is a very sticky issue in statistics and philosophy. One reason for this is – curiously – because there is little agreement as to exactly what the word causality means. You, thinking you had a pretty good idea of what the term 'cause' meant, might be surprised to learn that exactly what causation means is a matter of controversy. We will only cover the issue briefly, but a detailed review is given in Cook and Campbell (1979).

It is usually accepted that causation can be established if we satisfy three criteria: association, direction of influence and isolation. We will therefore consider what is meant by each of these in turn.

5.1.1 Association

One of the first things almost everyone learns in their introductory statistics classes is that 'correlation does not equal causation'. This is very true and we do not want anyone to think that we are saying that this is not the case, but unfortunately people are prone to taking this a little too far, and it turns out to be the only thing that they learned in their introductory statistics class. They then repeat this phrase, as if it were a mantra, responding to every supposition 'Ah yes, but correlation does not equal causation.' Whilst it cannot be doubted that 'correlation does not imply causation', it should also be remembered that 'causation *does imply* correlation'. If two variables are causally related, a change in one must produce a change in the other. Therefore, a statistical *association* (be it a regression coefficient or a correlation) is necessary but not sufficient to make a claim of causality.

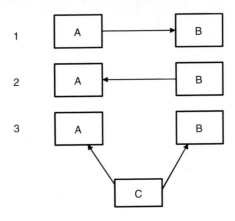

FIGURE 5.1 *Three possible directions of causality*

5.1.2 Direction of causality

If two variables (we will call them *A* and *B*) are associated, there are three possible reasons for this association. The associations are shown in Figure 5.1. First, it is possible that *A* is a cause of *B*. Second, *B* may be a cause of *A*. Third, another variable, which we will call *C*, is a cause of both *A* and *B*.

Therefore, for any correlation between two variables, it is not evident what underlying causal mechanism is giving rise to the observed association. In other words, we do not know the *direction of causality*. So how can we tell whether *A* causes *B*, or *B* causes *A*? (We will examine the situation where *A* and *B* are caused by another variable in the next section.) The answer is that we always expect the cause to come first in time, before the effect. If *A* causes *B*, a change in *A* should result in a change in *B* after a particular time period. So to find the correct direction of causality, we should be able to demonstrate temporal priority; that is, that changes in the dependent variable must be observed after a change in the independent variable, in other words *A* always precedes *B*.

Although it may appear self-evident that the cause must precede the effect, the actual time interval between cause and effect may vary widely depending on the variable in question. Consider the following examples. If we hypothesised that stopping smoking caused a reduction in heart disease, the appropriate time interval between cause (smoking cessation) and effect (reduction in heart disease) may be months or years. Measures of reduction in heart disease one week after the subject has quit smoking are unlikely to show any effect. Alternatively, if you attempt to measure the degree of pain caused by an electric shock, a time interval of milliseconds would be all that is required.

The notion of temporal priority is central to the basics of experimental design because the manipulation of the independent variable always precedes the measurement of the dependent variable. This temporal precedence cannot be observed in non-experimental or cross-sectional research where all data are usually collected at one point in time. In this case we have to rely on what Bollen (1989) called 'mental experiments' (p. 62). Mental experiments are decisions about the direction of causality based on theory, previous research and, in many situations, common sense. For example, if an association is found between gender and a person's level of depression it is clearly *not* sensible to claim that their level of depression 'causes' their gender.[1]

5.1.3 Isolation

To be certain that an independent variable, A, is a cause of a dependent variable, B, it is necessary to isolate the dependent variable (B) from all influences other than the hypothesised cause (A). This isolation is what experimentation attempts to achieve. Take as an example, an experiment designed to study the effects of temperature on memory. Participants could be randomly allocated to one of three conditions (room temperature 20°C, 30°C, and 40°C) and asked to complete a memory task. The experimenter would attempt to eliminate, or keep constant, all other variables that may influence memory such as time of day, light level and noise. If all these extraneous variables are eliminated then the only explanation for difference in memory performance across the conditions is temperature. The effect of temperature has been *isolated*. In practice experimentation can only approximate isolation, or achieve what is called *pseudo-isolation* (Bollen, 1989).

Whereas experimental control can be used to isolate the independent variable, in non-experimental research the independent variable(s) cannot be isolated. For example, if it is hypothesised that parental socio-economic status influences a child's school performance, there is no way that the researcher can physically control for all of the variables such as the child's IQ, parental IQ or parental income. But what the researcher can do is to isolate the effect of parental socio-economic status by statistically controlling for the influence of these potentially extraneous variables. From Chapter 2 on multiple regression, you should remember that a regression slope indicates the effect of the independent variable on the dependent variable *while holding constant the effect of all other independent variables in the equation*. Therefore regression models can be used to isolate the influence of an independent variable. It should be stressed that the successful isolation depends on the entry of all important extraneous variables as independent variables (more on this later).

5.1.4 The role of theory in determining causation

To tie together the three criteria discussed above we need to go beyond the statistical analysis — we need to involve theory. Fisher (cited in Cochran, 1965) was asked what could be done to move from correlational inferences to causal inferences. His reply was to 'Make your theories elaborate' (p. 252). What he meant by this was that the theoretical underpinnings of the analysis are vital in establishing a causal relationship – you cannot simply rely on statistical analysis. Consequently, you must have a thorough understanding of the potential mechanisms underlying the relationships between your variables. Many of those who repeat 'correlation does not imply causation' without thinking sufficiently about what it means do not understand the importance of making theories elaborate. Gould (1981) is highly critical of attempts to show causal relationships in psychological research using correlation-based methods: he lists examples of spurious relationships that arise, and shows how they are not causal. One example he gives is a very high correlation between his age and the price of petrol (gas), for the last 10 years. Gould says that it is tempting to leap to a causal conclusion, but to conclude that Gould's age affected the price of petrol would be a fallacy.

Atkinson, Atkinson, Smith, Bem and Nolan-Hoeksema (1996) give the example of the relationship between college grades and the number of hours (or amount of time) students spend on their college work. The correlation is small, but it is also negative. Students who spend more time studying tend to do slightly less well. This is a surprising and counter-intuitive result. Why does working harder seem to lead to lower grades? The answer is that it does not. No one said that working harder led to lower grades; what was said was that people who worked harder achieved lower grades.

We are not trying to argue that people should be encouraged to leap to causal conclusions every time they see a correlation. But if we were to follow the advice of Gould we would *never* draw causal conclusions. However, the mistake Gould (and many others) make is to pluck two correlated variables out of the air, and say 'Aha! These variables are related. Those dashed fool social scientists will go making causal inferences about them.' Well no, they will not, in the same way that a social scientist would not ask of an evolutionary theorist a question as naïve as that famously asked of Thomas Huxley by Bishop Wilberforce of whether he traced his descent from an ape on his grandfather's or his grandmother's side.[2]

When critics of causal analysis in psychological research find spurious correlations, such as that between age and the price of petrol, or the hours spent studying and final grades, they are making an analogy that psychologists wander around, plucking variables from the air and correlating them. Gould states 'In summary, most correlations are noncausal' (1982, p. 243). Of course this is true, but collecting data and correlating measures in the random fashion that this implies is not what psychological research involves. Collecting data is a costly, time-consuming process. Not only is it time consuming for the researcher, it is time consuming for the participants, who must give up their time to provide the data.

When psychologists collect data, they want to collect data that are both accurate and useful. Theory has a most important function in the successful use of regression and should not be underestimated. Psychologists think long and hard about their theories before even beginning to collect data. When Fisher said 'Make your theories elaborate', he meant that there should be theoretical input into the selection of the independent variables.

To use Gould's example, a researcher who was interested in examining the price of petrol would first think through all of the possible causes of changes in the price of petrol. They would consider factors like taxation, inflation, supply and demand of raw materials, and competition amongst retailers. They would have no reason to consider the age of a Harvard palaeontologist and popular science writer amongst the causes.

Similarly, if we consider the second example of students who spend more time on their college work performing less well, a researcher who was investigating students' grades (as opposed to making up some data that showed some examples for a textbook) would be foolish if they did not take some measure of ability. Students are usually aware of their ability and those students who have a very high level of natural ability know that they can pass the course without trying too hard (don't you just hate them?). Those students who are lacking in natural ability know that they have to try much harder to pass the course. (Those students who neither try hard, nor have a high level of natural ability, do not get to college in the first place.)

Abelson (1995) elaborates on this, referring to the 'method of signatures'. A series of findings, all related to correlations, can collectively be evidence of a causal process. This series of findings is referred to as the *signature* of that process. Abelson provides the example of smoking cigarettes being imputed as a cause of lung cancer. For many years, the evidence that smoking caused lung cancer was only correlational, and apologists (including Fisher himself) claimed that this was not therefore sufficient evidence to conclude that the link was causal.

Abelson states that if there is a causal link between smoking and cancer, in simple terms, it is because when tobacco smoke comes into contact with tissue, the smoke damages the tissue. There are a number of pieces of evidence that tie in with this hypothesis listed by Abelson:

1. The longer a person has smoked cigarettes, the greater the risk of cancer.
2. The more cigarettes a person smokes over a given time period, the greater the risk of cancer.
3. People who stop smoking have lower cancer rates than do those who keep smoking.
4. Smoker's cancers tend to occur in the lungs, and be of a particular type.
5. Smokers have elevated rates of other diseases.
6. People who smoke cigars or pipes, and do not usually inhale, have abnormally high rates of lip cancer.
7. Smokers of filter-tipped cigarettes have lower cancer rates than other cigarette smokers.
8. Non-smokers who live with smokers have elevated cancer rates. (Abelson, 1995: 183–184)

If there is some other variable that explains the non-causal relationship between smoking and lung cancer, then it would be very hard to explain why all of the above relationships occurred. Abelson goes beyond the presence of positive relationships to demonstrate a causal relationship. He also says that there should be no anomalous relationships: that is, there should be no relationships that are not explained by the hypothesised causal mechanism. For example, if the incidence of fallen arches of the feet were greater amongst smokers than non-smokers, this finding would be an anomalous result which would not fit in with the signature of the causal mechanism.

5.2 Sample size

(Or, to quote the title of a paper by Green (1991), 'How many subjects does it take to do a regression analysis?')

5.2.1 Why should we worry about sample sizes?

Until relatively recently there was very little said about sample sizes in psychological research. Introductory statistics textbooks make little mention of sample size as an issue, and we are aware of one that stated it was possible to have a sample that was too large to find an effect (this was in its first edition, and we are happy to report that this has now changed).

The bottom line regarding sample sizes is, or very nearly is, that bigger is better. Remember that the standard error of a mean is equal to:

$$\text{se}(\bar{x}) = \sqrt{\frac{\text{sd}^2}{n}}$$

Looking at this equation you can see that the larger the sample (N) is, the larger the denominator will be and therefore the standard error will be smaller. These smaller standard errors mean that your parameter estimates are more accurate. In terms of regression or correlation, an increase in sample size will also reduce the standard error thereby increasing the chance of finding a significant association. What this means in practice is that if your sample is too small you may not be able to detect associations that are present in the population and you might thereby reach the conclusion that variables are not related when in fact they are indeed related.[3] However, there are problems (although not statistical) associated with samples that are too large. If the sample was unnecessarily large, expense and time would have been wasted collecting more data than was necessary to carry out the study adequately, and some of the participants who gave their time would have done so unnecessarily. Ethical review boards and research funding councils have started to pay a great deal more attention to issues of sample size than in the past. Ethical review boards have become aware that research involving human participants involves some degree of inconvenience: participants donate their time in the hope that they may be doing some good. If a study did not have a sufficiently large sample, those people would have given their time for research that was not of high enough quality to be useful. Similarly, if a sample were too large, then some of the participants who gave their time would have done so unnecessarily. Bodies which fund research have an interest in knowing that the research is to be as efficient and as cost effective as possible.

There are two different ways to go about determining an appropriate sample size, the use of rules of thumb or using power analysis. Rules of thumb represent simple rules that suggest minimum sample sizes. Power analysis is an objective statistical technique (or a collection of techniques) for determining appropriate sample sizes for any study or experiment. In an ideal world we would all use power analysis all of the time. However, power analysis can be complex and a little daunting for the beginner. We will examine both approaches below.

5.2.2 Rules of thumb

Most rules of thumb tend to be very simple. For example, it has been suggested that that there should be more than 100 participants, or that there should be (ideally) at least 20 participants per independent variable in a regression analysis. Green (1991) proposed a method for determining a minimum sample size to test the R^2 of a regression model. He suggested that the minimum sample should be greater than $50 + 8k$, where k is equal to the number of independent variables. If you want to carry out significance tests on regression slopes, the size should be greater than $104 + k$, where k is again equal to the number of independent variables. If you want to test the R^2 and slopes you should calculate the minimum sample size for both values and use the largest.

The problem with such rules of thumb is that they do not take into account issues such as the expected effect size or the desired power of the test. Because they do not take these into account, such rules lack generality and may even mislead. For this reason power analysis is the preferred option.

5.2.3 Power analysis

To use power analysis to calculate the sample size requirement we need the following pieces of information: the significance level being used (this is called alpha), the effect size, and the appropriate level of power. These are described as follows:

The value of alpha: The value of alpha is the level of significance that we are going to use as the criterion for determining whether or not we have a significant effect. By convention, this value is usually set to be 0.05. The larger the value of alpha, the more chance we have of finding a significant result, but the chance of finding a significant result by having a larger alpha must be traded off against the probability of getting a spurious result (a type I error). A type I error occurs when we find a significant association in the sample that is not present in the population. The type I error rate, that is the probability of making a type I error, is equal to the value that we choose for alpha. That the type I error rate is equal to alpha is a fact that can confuse many students (and not a small number of lecturers and researchers) so we will make it doubly explicit: the probability of committing a type I error has nothing to do with your sample size (or the test type, if the assumptions are satisfied). It is always equal to the cut-off value — alpha — that you choose to use for your significance test.

The size of the effect in the population that we would be interested in: If we measured the entire population from which we have randomly selected our sample, the size of the correlation or regression coefficient we would find is known as *the effect size*. In multiple regression, the effect size is equal to R^2. The larger the effect size, the greater the chance we have of finding it. However, if the effect size is sufficiently small, then finding it would not be useful. According to Kraemer and Thiemann (1987), the size of the effect can be determined in one of three ways.

First, the effect could be based on substantive knowledge. What value of R^2 would be sufficiently high to be useful? It depends on the effect that is being investigated. We would want to know about some effects, even if they were very small. An example of this is the potential danger of living near high-voltage power lines. The effect of living near power lines may be very small — it may only be noticeable in 1 in 500 people who live near such power lines. From a practical point of view, in a large population, 1 in 500 adds up to an awfully large number of people, and from a personal point of view, even if the chances were that low, and you lived near such a power line, you would want to know about any effects. From a theoretical point of view such an effect would be interesting, because it would tell us something about the biological process involved, which we knew nothing about. Therefore, if we were carrying out research investigating such an effect, we would want to make sure that we had sufficient power to detect a very small effect. Other effects would not be interesting unless they were very large. If the research was investigating the teaching of statistics using two different statistical software programs, to see which was more appropriate, we would not be very interested in an effect in which 1 extra student in 500 would pass their statistics course. If the effect of changing a software package meant that only 1 more student in 500 was going to pass, then which software package was used would depend on factors other than this one. It is unfortunate for that one student, but a student would never say 'Well, I was going to do psychology at Poppleton University, but they use SPSS, so I chose the University of Uttoxeter, who of course use Minitab.'

The second way of determining the expected effect size is to base an estimate on previous research. To do this, we would see what sort of effect sizes other researchers studying similar fields have found, and we could use their results to get an estimate of the effect size we would be likely to find.

The third possibility for determining expected effect size is to use conventions. Cohen (1988) has defined small, medium and large effect sizes for values of R^2. He defined a small value as $R^2 = 0.02$, a medium size as $R^2 = 0.13$ and a large size as $R^2 = 0.26$. These form useful conventions and can guide you, if you know approximately how strong the effect is likely to be.

Power: An appropriate level of power must be selected. *Power* is the probability of finding a result given that the effect does exist in the population.

By convention (from Cohen, 1988), power is set to 0.80. This gives an 80% chance of finding a significant result if there is an effect of the specified size in the population from which the sample is taken. The probability of a type II error is equal to 1 − power, which is equal to 20%. By using these conventional values of alpha = 0.05 and power = 0.80, we are (implicitly) stating that a type I error is four times more serious than a type II error.

Once we have all the information we need, the power, the effect size and alpha, we can calculate the number of participants required. However, the calculations required for power analysis are complex, and so are rarely attempted by hand. Instead, tables of pre-calculated values can be consulted (see e.g. Cohen, 1988), or computer programs can be used (GPower, by Faul and Erdfelder (1992), is a free program that can be used for many types of power analysis). Here we will use charts to make our sample size estimates. The data that were used to draw the following charts were generated using GPower. We do not recommend using these charts for accurate power analysis, but they can be used to draw some conclusions about your sample sizes, and the sample sizes of research that you encounter. The value for alpha is always set to be 0.05, and we always follow Cohen's conventions for effect sizes. In all of the following graphs, the thick solid line represents the large effect size, the thick dashed line the medium effect size, and the thin solid line the small effect size.

Let us suppose that we have planned a study with two independent variables. We expect a large effect size, and use 0.80 as an acceptable level of power. To use the chart, you first find the chart for the correct number of variables (in this case two). Then on the y-axis locate 0.80, which is the power. From this point move across the chart until you reach the solid line, which represents a large effect size. The sample size can be read off the corresponding point on the x-axis. In this case, the sample size required is about 32 (the scale is logarithmic to spread the lower end out further). You should check this to make sure that you agree with us.

Let us now use an example that is more like the typical study in psychology. Let us suppose that we have planned a study with four independent variables (a modest number compared with many published research papers); we expect a small effect size, and still use 0.80 as an acceptable level of power. The sample size required is now 600.

You can see from the charts that when maintaining power at 0.80 the sample size requirement becomes greater first as the number of independent variables increases and second when the effect size is small. This means that the sample size requirement for many psychological studies will be larger than the sample sizes that are generally used and larger than many researchers expect. Power analysis calculations can often make us feel that our sample sizes are woefully inadequate (and we are afraid that these feelings sometimes occur because sample sizes are woefully inadequate). A number of surveys of the power of published psychological research also back up the view that researchers need to think more about power (e.g. Cohen, 1962; Sedlmeier & Gigerenzer, 1989; Clark-Carter, 1994).

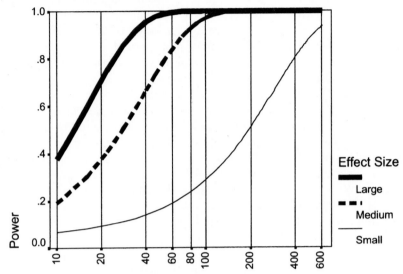

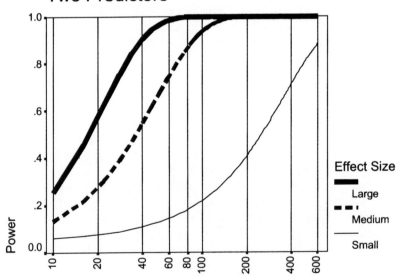

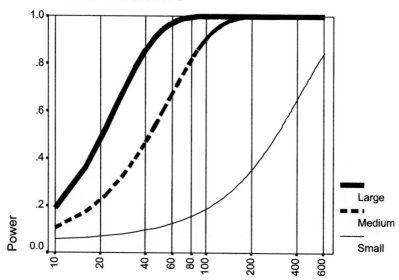

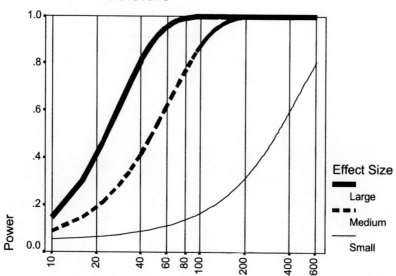

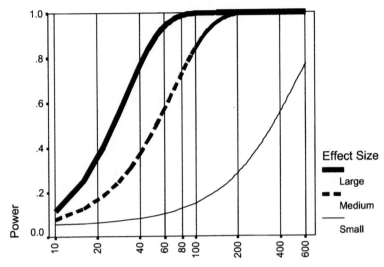

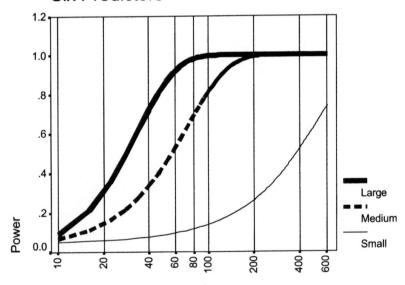

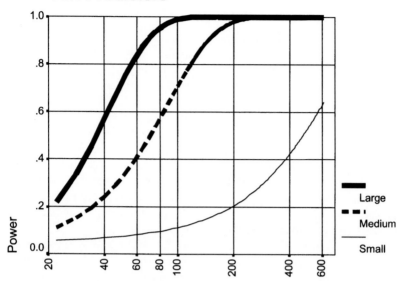

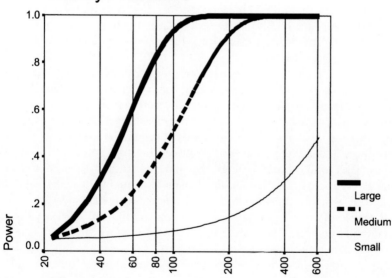

5.3 Collinearity

5.3.1 What is collinearity?

The next issue that we will examine is collinearity, sometimes referred to as multicollinearity (although some argue that these terms have slightly different meanings, most people, including us, use them interchangeably). Collinearity refers to the size of correlations among the independent variables in a regression calculation (literally co-linearity, lying along the same line).

Collinearity happens because two (or more) independent variables correlate. This means that it is difficult for the regression calculation to determine which of them is actually the important one of the two; it could be either, so we have increased uncertainty (standard errors) and inaccuracy (slope coefficients). When there is complete collinearity, we have no way of knowing which of the two (or more) variables are important, and therefore a solution cannot be found. This sounds like a difficult statistical algorithm, but it is what we all do when faced with a similar sort of problem – and we do not find it difficult, although it seems like a difficult concept, so we will look at a simple example.

It is a commonly held belief that red cars are more likely to be involved in accidents than cars of any other colour. If this is true, it may be because there is something about driving a red car that makes one more likely to be involved in an accident. (Drive more quickly? Pay less attention?) Alternatively, it may be because the type of person that decides to buy a red car is more likely to be the sort of person that is more likely to be involved in an accident. However, people driving red cars are likely to be the sort of people that like red cars, so the two variables are highly correlated. It is difficult to work out, using a statistical technique, which of these two is the factor that determines how likely a car is to have an accident.

There is a difference in the behaviour of males and females; few people would argue with this fact. There has long been a debate in the social sciences about why males and females behave differently. Is it because of genetic biological differences between them, or is it because of their life experiences and upbringing? Using statistical methods it would be impossible to know, because there is complete collinearity between the two potential independent variables. Everyone who is biologically male is brought up and treated as a male. Everyone who is biologically female is brought up and treated as a female. It is possible that all of the differences between males and females occur because of their upbringing, or it could be that all of the differences between males and females are due to genetic differences. But there is no way of knowing using purely statistical techniques.

These examples show how we are able to think about problems and find that when two variables covary (change together) we cannot decide which is important in determining the outcome. The regression calculations simply take account of this uncertainty, and call it a larger standard error.

In more formal terms, when the correlation between two independent variables is one (or very close to one) or the multiple correlation between any independent variable is one (or very close to one), this is referred to as perfect or complete collinearity (or multicollinearity). It is an assumption of regression that perfect collinearity is not present, and if perfect collinearity does occur, most statistical packages will stop and produce an error message. However, perfect collinearity is a very infrequent occurrence with real data, unless a data entry or manipulation error of some sort has been made. If this does occur it is most likely (in our experience) that two or more independent variables have been summed to create an additional variable that is then also used as an independent variable. Many measures in psychology and the social sciences comprise a number of subscales, which are summed to create a total score. For example, an intelligence test may contain abstract reasoning, verbal reasoning and mathematical reasoning and these three subscores are then summed to create a total score. If all four of these variables, the three subscores *and* the total, are entered as independent variables into a regression analysis, the assumption of lack of perfect collinearity will be violated. If we used the total score as a dependent variable in a regression equation, and the subscores as an independent variable, we would find that R^2 would be equal to one — we could make a perfect prediction as to the value of the total score if we knew all of the subscale scores (we should, because we just created it).[4] In addition, if we know any of the three scores, we can work out the value of the fourth:

$$\text{Score } 1 = \text{total} - (\text{score } 2 + \text{score } 3)$$

In this example, the fourth variable is said to be redundant: it adds no more information that we did not already know occurs. The diagnostic techniques described below will help you to isolate the problem.

A much more common occurrence is for collinearity to be high enough to cause some problems, but not actually high enough to violate the assumptions of regression. In the next section, we will look at how this problem of high collinearity may be detected.

5.3.2 Detecting collinearity

In this example we will take dataset 2.1 from Chapter 2 (the one that contained information about the number of books that students had read, the number of lectures they attended and the grades they achieved in a statistics course). We will add to this dataset (Table 5.1) a fourth variable, called `late`, the number of lectures to which students turned up late.

If we reanalyse this dataset, including the variable `late` as an independent variable, we find that $R = 0.621$, $R^2 = 0.386$, $F = 7.54$, $df = 3, 36$, $p < 0.005$; thus the overall equation is significant and predicts a great deal of the variance in grades. However, when we examine the significance tests for the slopes (Table 5.2), we might be surprised at the result.

We find that none of the values of the regression coefficients themselves are significant (at the 0.05 level). This may seem curious, and can confuse you

TABLE 5.1 Dataset 5.1

books	attend	grade	late
0	9	45	7
1	15	57	8
0	10	45	9
2	16	51	4
4	10	65	2
4	20	88	1
1	11	44	9
4	20	87	0
3	15	89	3
0	15	59	8
2	8	66	8
1	13	65	7
4	18	56	5
1	10	47	8
0	8	66	8
1	10	41	10
3	16	56	7
0	11	37	4
1	19	45	4
4	12	58	4
4	11	47	7
0	19	64	7
2	15	97	4
3	15	55	3
1	20	51	6
0	6	61	5
3	15	69	3
3	19	79	6
2	14	71	7
2	13	62	6
3	17	87	5
2	20	54	2
2	11	43	8
3	20	92	3
4	20	83	2
4	20	94	0
3	9	60	8
1	8	56	9
2	16	88	4
0	10	62	7

TABLE 5.2

	Slope (*b*)	Std error of slope	Standardised slope (beta)	*t*	Sig.
Constant	60.941	14.899		4.090	0.000
books	2.318	1.942	0.199	1.193	0.240
attend	0.702	0.652	0.180	1.078	0.288
late	−2.188	1.195	−0.347	−1.831	0.075

when you first start to use regression analysis, but when you find that the regression coefficients are not significant, when the overall equation is significant, you should suspect that collinearity may have played a part. We said earlier that collinearity increases the uncertainty around the parameter estimates, and therefore increases the standard errors. What may have happened in this case is that the regression equation 'knows' that a high proportion of the variance can be explained by the independent variables, but it does not know what size parameter estimates to assign to which independent variable. Just as in the previous example we know that there is some variance in the behaviour of males and females (the overall equation is significant), but we do not know what the parameter estimates for 'environment' and 'genetics' are.

If you suspect collinearity problems, there are several methods of determining the severity of the problem. Three such methods are discussed below.

The first method of assessing collinearity is simply to inspect visually the matrix of correlations amongst the independent variables. This is shown in Table 5.3.

TABLE 5.3 Correlation matrix of independent variables

	books	attend	late
books	1.000	0.444	−0.615
attend	0.444	1.000	−0.617
late	−0.615	−0.617	1.000

It is clear that the variable late correlates fairly highly with the other two variables and this high correlation is a clue that collinearity may be a problem. Be warned, however, that low correlations do not indicate that there is no problem, for it is the multiple correlations that matter, not the bivariate correlations. The correlation tells us how much variance two variables share, but this figure does not determine whether collinearity is a problem. The value we need to know is the proportion of variance in each independent variable, which is shared by all of the other independent variables. The way to find out how much variance a variable shares with other variables is to carry out a multiple regression analysis, using each independent variable in turn as a dependent variable, and all of the other independent variables as independent variables.

The results of this analysis are presented in Table 5.4. When the dependent variable is either books or attend, R^2 is approximately 0.35, meaning that

TABLE 5.4

IV	DVs	R	R^2
books	attend, late	0.620	0.351
attend	books, late	0.622	0.354
late	attend, books	0.725	0.500

IV, independent variable; DV, dependent variable

slightly more than one-third of the variance in each of those variables is shared with the other independent variables. However, when late is used as an independent variable we find an R^2 of 0.500, meaning that half of the variance in late is shared with the other independent variables. Effectively this means that half of the information that late tells us, we already knew, and so half of the information was redundant.

Doing a separate regression analysis for every independent variable would become rather tedious, but fortunately most statistical analysis software will give two other diagnostic statistics to help you to diagnose collinearity. These are the tolerance and the variance inflation factor (VIF).

Tolerance is a very slight extension of R^2; the tolerance of an independent variable is the extent to which that independent variable cannot be predicted by the other independent variables. Tolerance for a variable is calculated as $1 - R^2$, where the variable being considered is used as the dependent variable in a regression analysis and all other variables are used as independent variables. If there are only two independent variables in the regression analysis, the value of R (the multiple correlation) will be equal to the value of r (the bivariate correlation), and the tolerance can therefore be calculated from the bivariate correlation matrix. Tolerance varies between zero and one. A tolerance value of 0 for a variable means that it is completely predictable from the other independent variables, and that therefore there is perfect collinearity. If a variable has a tolerance value of 1, this means that the variable is completely uncorrelated with the other independent variables.

The variance inflation factor (VIF) is closely related to the tolerance. When there are more than two independent variables the VIF is calculated using the following formula:

$$\text{VIF} = \frac{1}{\text{tolerance}}$$

Although it may seem superfluous, the VIF is useful because it relates to the amount that the standard error of the variable has been increased because of collinearity. The increase in standard error is equal to the square root of the VIF: when the VIF is equal to four the standard error is doubled ($\sqrt{4} = 2$), and so four is often used as an arbitrary cut-off to determine when collinearity has become too serious.

The tables presented below show the output from a regression analysis of dataset 5.1. The tables include the estimates of tolerance and VIF. In Table 5.5 attend and books were used as independent variables. The tolerance values of these variables are equal, at 0.803, and VIF is 1.245. Both the VIF and the tolerance values are acceptable, as we require a tolerance value close to one and a VIF value less than two. If we examine Table 5.3 we can see that the correlation between the two variables is equal to 0.444. You can confirm that $1 - 0.444^2$ is equal to 0.803, thus giving the tolerance value (and showing why it is unnecessary to consider the tolerance when we have two independent variables only).

TABLE 5.5 *Collinearity diagnostics for two independent variables*

	Slope (b)	Std error	Beta	Sig.	Tolerance	VIF
attend	1.283	0.587	0.329	0.035	0.803	1.245
books	4.037	1.753	0.346	0.027	0.803	1.245

Table 5.6 shows the results from the analysis when we include the variable late as an independent variable. We can see that the tolerance value for the attend and books variables has decreased, but the tolerance value for late is considerably lower and VIF is greater than two, alerting us to the possibility of collinearity.

TABLE 5.6 *Collinearity diagnostics for three independent variables*

	B	Std error	Beta	Sig.	Tolerance	VIF
late	−2.188	1.195	−0.347	0.075	0.474	2.108
attend	0.702	0.652	0.180	0.288	0.613	1.632
books	2.318	1.942	0.199	0.240	0.615	1.625

Therefore, we have methods of detecting collinearity, but what do you do if you find that collinearity is a problem with your dataset? We will attempt to answer this question in the next section.

5.3.3 Dealing with collinearity

Fox (1991) stated that there is 'no quick fix' (p. 13) to the problem of collinearity in regression. In fact, the best thing to do if collinearity is a serious problem is to discard the old data and go and collect new data, which avoids the problem in some way. This advice is usually not useful or practical, so we suggest some other ways of dealing with the problem.

5.3.3.1 Collect more data

We just said that we would not suggest collecting new data, and we are not. We are not suggesting collecting *new* data, we are suggesting collecting *more* data. Although this may be the easiest solution to describe, it is the hardest to implement. In addition, it is not much of an improvement on collecting new data. Collinearity causes the standard errors to increase in size. We saw in Chapter 1 that larger samples have smaller standard errors, so a larger dataset will, in some way, make up for some of the effects of collinearity. Of course collecting more data would not help when *perfect* collinearity is present.

5.3.3.2 Remove or combine variables

If the variables are highly correlated this implies that they are measuring similar constructs and that the information in one of those variables may be, at least partially, redundant. Therefore one solution is to remove one of the variables or to combine the variables in some way.

If we were using parents' educational level as an independent variable, we would have two scores, mother's educational level and father's educational level. In this case, we would probably find collinearity to be a problem, as people tend to pair with people of a similar educational level. In such a case, removing one of the variables is a solution – we would keep whichever variable had the greater theoretical relevance for our model, either mother's education or father's education, but not both. Alternatively we could combine the scores on the two variables and use this new 'parental education level' variable as an independent variable. Similarly, if we were looking to see if individual module grades in year 1 of a degree course were able to explain overall degree classification we would probably find collinearity to be a problem. People who do well in one module tend to do well in other modules, and those who do poorly in one tend to do poorly in others. If this is the case, combining the grades by taking an average, or finding a total score of all the modules, may be the best thing to do.

If you are dealing with a large number of independent variables, from a questionnaire for example, a more technical way of reducing the number of variables is to use principal components analysis (PCA). This procedure is similar to factor analysis in that it groups the original variables into a smaller number of uncorrelated 'factors' (strictly, components). The resulting factors can be used as independent variables in subsequent regression models without producing collinearity problems, as the factors are by definition uncorrelated. A full discussion of PCA is beyond the scope of this text, however we believe that Kline (1994) is a very good starting place for more information.

5.3.3.3 Stepwise entry
Stepwise entry is a form of hierarchical regression, as discussed in Chapter 2. It can be used when collinearity is a problem to select variables for analysis. In Chapter 2, we listed the serious problems with using stepwise regression. For the reasons that we stated there, we strongly advise against its use. The only time we might temper this advice is when analysis is being carried out *purely* to predict a dependent variable, and when no consideration is being taken in the underlying theoretical links. (Although we might ask why you were interested in such a thing.)

5.3.3.4 Ridge regression
When collinearity is so high that the regression procedure cannot continue, a possible solution that has been proposed is ridge regression. Ridge regression is a complex procedure, which is beyond the scope of this book. It is also not easily implemented within most statistical packages, and it can be difficult to interpret and for these reasons it is rarely used. For more information see Pagel and Lunneborg (1985) or Draper and Smith (1981).

5.4 Measurement error

Thorndike (1917) wrote, 'any measure is a compound of fact and errors which the instrument will surely make' (p. 207). This measurement error has the consequence of attenuating (reducing) correlations between two variables,[5] and the correlation between two variables should be distinguished from the constructs or conceptual variables to which they relate (Cohen and Cohen, 1983).

The reliability of any measure (r_{xx}) can be defined as the correlation between the variable as measured and another equivalent measure of the same variable. The square root of the reliability, in classical test theory, is the correlation between the true score and the score obtained by the measuring instrument.

The correlation between two scores measured at less than perfect reliability will be lower than the true correlation between the scores. Figure 5.2 illustrates the attenuation caused by lack of reliability. We measure the variables x and y, but we are not really interested in x and y per se, we are interested in the underlying psychological construct that x and y are measuring. If you fill in a questionnaire that assesses your personality, we are only interested in the answers you give, because they tell us something about your personality – they are indirect (and imperfect) measures of personality. The true measures of personality, that we are interested in are represented by x^* and y^*. The reliability of each of the measures is represented by r_{xx} and r_{yy}. The correlation between the two measured variables is shown as r_{xy}, but this is not what we really want to know. We are interested in the correlation between the true scores – $r_{x^*y^*}$. That correlation is the correlation between the psychological constructs, and that is what we really want to know.

We can also represent this as an equation:

$$r_{xy} = r_{x^*y^*} \sqrt{r_{xx} r_{yy}}$$

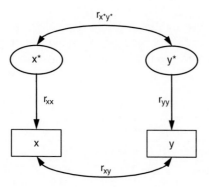

FIGURE 5.2 *Path diagram representation of estimated and actual correlations*

It is possible, therefore, to make a correction based on reliability, and given by Cohen and Cohen (1983), referred to as the *attenuation-corrected correlation coefficient* and shown in the following equation:

$$r_{x*}r_{y*} = \frac{r_{xy}}{\sqrt{r_{xx}r_{yy}}}$$

In conclusion, unreliability can have a dramatic effect on correlations. If, for example, the true correlation between two scores is 0.90 (a very high correlation), but the reliability of the two measures is 0.60 (a not unusually low reliability), the correlation between the measured items will be lowered to 0.54, a considerable drop.

Shevlin (1995a, 1995b) has demonstrated in a series of Monte Carlo simulation studies how failure to account for reliability can cause type II error rates to rise. Cohen and Cohen (1983; see also Cohen, Cohen, Teresi, Marchi and Velez, 1990) recommend the cautious use of the attenuation correction, or else the simple acknowledgement that correlations may be reduced because of unreliability, stating:

> unreliability in variables is a sufficient reason for *low* correlations; it can not cause correlations to be spuriously high. (Cohen and Cohen, 1983: 7)

Attenuation correction should be employed with extreme caution, as it makes the assumption that all error in the measurement of the variables is random and uncorrelated. It is very rare that this assumption holds true, and there is rarely any attempt to test it. Employing a structural equation modelling approach (see Chapter 8) would use attenuation correction as part of the analysis, but also makes explicit the assumption about error being random and uncorrelated, and ensures that it is tested.

In a multivariate situation the position becomes more complex. Bollen (1989) provides an example based on data collected from 108 areas of the USA, and a model suggested by Lave and Seskin (1977). In the model there are seven independent variables and one dependent variable. It is assumed that all variables except the dependent variable are measured with perfect reliability. Multiple regression was carried out, with reliabilities inserted into the dependent variable of 1.0, 0.9, 0.7 and 0.5. The results are shown in Table 5.7.

TABLE 5.7 *The effects of unreliability on the beta weights from a regression (adapted from Bollen, 1989)*

Reliability	IV_1	IV_2	IV_3	IV_4	IV_5	IV_6	IV_7	R^2
1.0	0.107	0.090	0.064	1.008	0.370	−0.063	−0.076	0.839
0.9	0.123	0.086	0.060	1.003	0.369	−0.062	−0.072	0.840
0.7	0.173	0.072	0.047	0.986	0.363	−0.056	−0.056	0.845
0.5	0.291	0.039	0.017	0.947	0.350	−0.044	−0.020	0.855

It can be seen in the table that as the reliability is lowered, the regression weights alter, but in a far from predictable fashion. The loadings from IV_1 increase quite dramatically as the reliability falls, whereas the loadings from IV_2, IV_3 and IV_7 (in absolute terms) fall, and IV_5 and IV_6 remain relatively unaffected.

In the situation where more than one variable is measured imprecisely, the effects of measurement error become more unpredictable.

Notes

1 It may seem strange that this needs to be stated, but we have read claims by students, brought up on 'correlation does not imply causation', that a sex difference does not mean that sex is the cause of the difference. Is something 'causing' sex?

2 Rather appropriately, when this question was asked, Darwin took the stage and said 'A man has no reason to be ashamed of having an ape for his grandfather. If there were an ancestor whom I should feel shame in recalling, it would be a man of restless and versatile intellect, who, not content with success in his own sphere of activity, plunges into scientific questions with which he has no real acquaintance, only to obscure them by an aimless rhetoric, and distract the attention of his hearers from the point of issue by eloquent digressions.'

3 It may be tempting, and common, to think in this way. However, you have probably been told by your statistics tutors that you do not accept the null hypothesis, rather you should fail to reject the null hypothesis (not quite the same thing).

4 This can be a useful technique if you do not know how a total score was calculated, because you have forgotten, or you are analysing someone else's data.

5 Throughout this discussion, when we say reduce, we really mean 'reduce in absolute value'. Negative correlations will be increased.

Further reading

Causality is covered in Bollen (1989) *Structural equations with latent variables*, Chapter 3, Maxwell and Delaney (1990) *Designing experiments and analysing data: a model comparison perspective*, Chapter 1, and Cook and Campbell (1979) *Quasi-experimentation*.

Power analysis is covered in general in Cohen (1988) *Statistical power analysis for the behavioral sciences*, and Kraemer and Thiemann (1987) *How many subjects? Statistical power analysis in research*. Specifically regarding regression analysis, see Cohen and Cohen (1983) *Applied multiple regression/correlation analysis for the behavioral sciences* (2nd ed.), Chapter 3.

Collinearity is covered in Fox (1991) *Regression diagnostics*, Chapter 3, and Pedhazur (1997) *Multiple regression in behavioral research: explanation and prediction*, Chapter 7.

Measurement error is covered very thoroughly in Bollen (1989) *Structural equations with latent variables*, Chapter 5, and more briefly in Berry (1993) *Understanding regression assumptions*, Chapter 5.

PART III

I NEED TO KNOW MORE OF THE THINGS THAT REGRESSION CAN DO

6 Non-linear and logistic regression

6.1 Non-linear regression

6.1.1 Linear and curvilinear relationships

So far in this book, we have looked exclusively at *linear* regression analysis. A linear relationship occurs when an increase in the independent variable is associated with a constant increase in the dependent variable. A linear model is simple to describe, but there are many circumstances in the social sciences where a linear model is not suitable for describing the situation. One classic example is the Yerkes–Dodson Law of Arousal (Yerkes and Dodson, 1908), which describes the relationship between stress and performance. The famous inverted-U is shown in Figure 6.1. When arousal is very low, performance is poor (when you are tired and don't care). As arousal increases, so performance increases, but only up to a point. When arousal increases beyond a particular point performance begins to deteriorate (something most of us are familiar with in exams).

If we tried to model this relationship using linear regression, we would find that the analysis suggested that there was no relationship between the variables when clearly this is incorrect, there is a relationship.

Many relationships will be seen to be non-linear if we extend them far enough. In the example we looked at in Chapter 1, we assumed that the effect of reading a textbook was a constant increase in grades, but we know there are limits to this effect. The improvement effect of reading the first textbook is going to be very different from the improvement effect of reading the 10th textbook, which in turn will be different from reading the 100th textbook. This type of situation, which would be illustrated by a curve whose shape would indicate that the effect would become less, is very common in many areas, as anyone who has attempted to improve at a sport, learn a musical instrument, lose weight or write an essay can testify. The graph in Figure 6.2

NON-LINEAR AND LOGISTIC REGRESSION   137

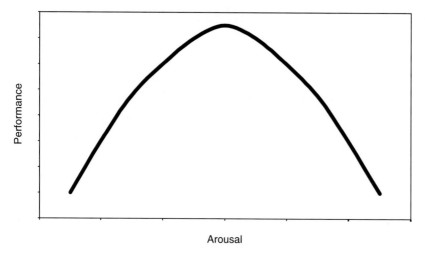

FIGURE 6.1   *The Yerkes–Dodson law*

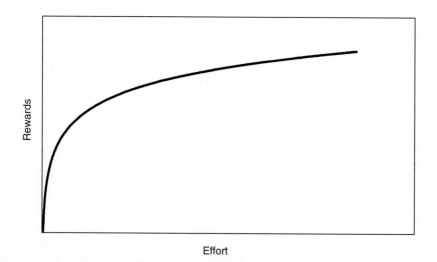

FIGURE 6.2   *Efforts vs rewards – a curvilinear relationship*

shows the general shape of this relationship: at lower values of effort, an increase in effort gives a great deal more reward, but at higher values of effort, the increase in reward is much less.

6.1.2 Generating a curve

To fit a curvilinear model we need to use a different type of equation. Where we previously used the formula

$$y = bx + c$$

this meant that our predicted value of y is equal to the slope (b) multiplied by x, plus a constant (c). This equation can always be represented by a straight line on a graph, and hence is referred to as a linear relationship. Sometimes we have prior reason to believe that the relationship between two variables will not be linear, or it may be clear from examining a scattergraph that there is a non-linear relationship between two variables. Where we think that the relationship between variables may not be linear, we need to fit a slightly different type of equation, known as a non-linear or a curvilinear regression. To fit this type of model, we need to do something to turn a straight line (the slope) into a curved line. If we can take a straight line and curve it, we can do the opposite, which is to take a curved line and straighten it. In addition, if we have a curved line that has been straightened, we can do on this straight line all the processes that we have been using in linear regression, just as we have been doing all along.

6.1.2.1 Linear transformations

A transformation is something we do to every datum in a variable (datum is the singular of data). If we take the series of integers from 1 to 20 (which we could call x), multiply each number by a value (to create the variable that we will call y) and plot x against y on a scattergraph, they will always end up as a straight line. Similarly, if we add a number to every value of x, the relationship will end up as a straight line (Figure 6.3).

By doing these types of transformations, we may change the constant and the unstandardised slope coefficients in a regression equation, but we will not alter the standardised slope coefficients, R^2 or the probability values associated with any of these values.

We need to do something to x, rather than simply increase it by multiplying by something or adding something, if we are to make the straight line turn into a curved line. Several functions are available that will do this and we will examine some of the more common ones.

6.1.2.2 The quadratic function

To create a quadratic function we square the value of the independent variable and by squaring all these values, we create a curved line. Figure 6.4 shows the plot of x against x^2. We find that at low values of x the curve is flat, and then at higher values of x it gradually curves upwards, becoming steeper and steeper. This happens because when we square a series of numbers the smaller numbers increase a little, but the larger numbers increase a great deal. If participant A scores 2, and we square their score, they now score 4, twice as large. If participant B scores 4, when we square this number it becomes 16, which is four times larger. In the raw score, participant B scored twice what participant A scored, and in the transformed score participant B scored four times what A scored. The effect becomes more pronounced as numbers get higher. Participant C, who originally scored 12, has a score six times larger than participant A, and three times larger than participant B. If

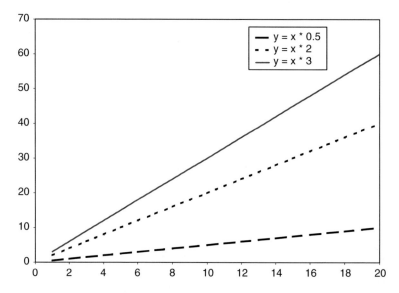

FIGURE 6.3 *Graph of* x *plotted against* y = x × 0.5, y = x × 2 *and* y = x × 3

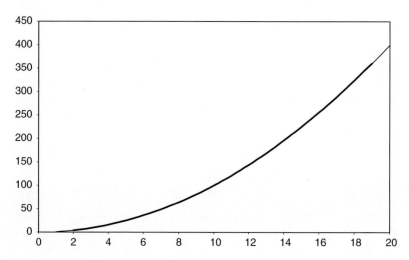

FIGURE 6.4 x *plotted against* x^2

we square participant C's score, they score 144. They are now scoring 36 times participant A's score (as opposed to six times previously) and nine times participant B's score (as opposed to three times previously).

Previously, our linear regression equation looked like:

$$y = b_1 x_1 + c$$

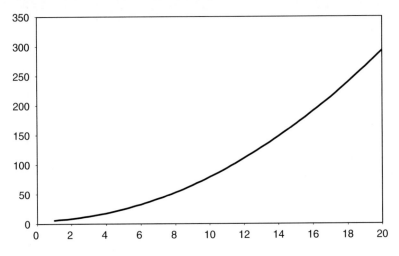

FIGURE 6.5 *Graph of* $y = 0.4 \times x_1 + 0.7 \times x_1^2 + 5$

where b_1 and c were estimated. This equation always gave a straight line. Now we use the quadratic equation:

$$y = b_1 x_1 + b_2 x_1^2 + c$$

which will give a curved line. You may notice that this equation is very similar to those presented in Chapter 2 for multiple regression. In this case, the slope for the second variable (x^2) does not relate to a second independent variable but to the variable we have created by squaring the values of the original x variable. We have substituted some values into the equation and plotted values of x. The graph given by the equation:

$$y = 0.4 \times x_1 + 0.7 \times x_1^2 + 5$$

is shown in Figure 6.5.

A second use of the quadratic function is to subtract the quadratic from a constant, which creates a line that decreases slowly at first and then rapidly increases in its rate of decrease. Figure 6.6 shows a slope of $y = 1000 - 2x^2$.

6.1.2.3 The cubic function

The quadratic function contained x^2 as an independent variable, which allowed a line to change direction, but only once – it could start sloping gently, and then get steeper. The cubic function, in contrast, allows a line on a graph to change direction twice.

Cubic regression equations take the form:

$$y = b_1 x_1 + b_2 x_1^2 + b_3 x_1^3 + c$$

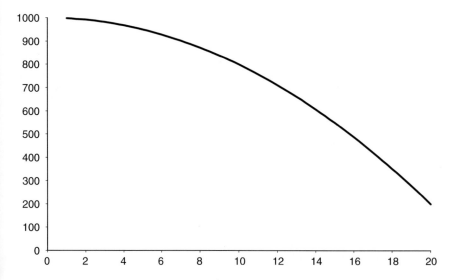

FIGURE 6.6 A curve of $y = 1000 - 2x^2$

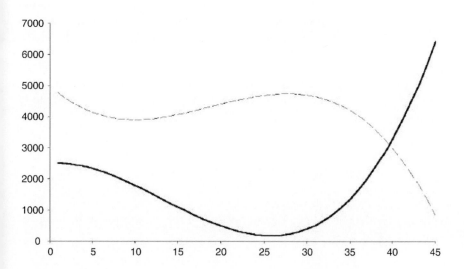

FIGURE 6.7 Cubic functions of $y = -250 \times x + 17 \times x^2 - 0.3 \times x^3 + 5000$ (dashed line) and $y = 20x - 12x^2 + 0.3x^3 + 2500$ (solid line)

By altering the values of the three slopes (the *b*-values), a range of curves can be produced. Figure 6.7 shows two curves that are produced from two different equations.

Generating the data is done in the same way as with the quadratic curves, but in this case we have a third independent variable, which is the cube of the original value of *x*.

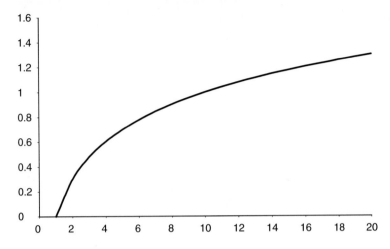

FIGURE 6.8 *Graph showing* $y = \log(x)$

6.1.2.4 The log function

The logarithm to base 10 (or log) of any value is the number that you would need to raise 10 to the power of, to give the original value. Thus the log of 10 is 1, because $10^1 = 10$, the log of 100 is 2, because $10^2 = 100$, and the log of 1000 is 3, because 10^3 is 1000. Where the square function has the effect of stretching out the higher values of x, the log function has the opposite effect of squashing together the higher values. This leads to a curve that starts sloping steeply and then gradually levels off. Figure 6.8 shows an example of such a graph.

6.1.2.5 The inverse function

The final curvilinear function that we shall examine in this section is the inverse function. The inverse is calculated as $y = 1/x$. This has the effect of making small values (less than one) into large values and large values (greater than one) into small values. Figure 6.9 shows the curve produced using the values 0.1 to 3.0.

6.1.3 Carrying out non-linear regression

Carrying out a non-linear regression analysis is as simple as transforming the independent variable and adding it to the equation. There are a few points to be borne in mind.

6.1.3.1 To add a constant, or not to add a constant

When we are carrying out linear regression analysis, we can manipulate the variables in several ways, and, as long as we do the same thing to every case,

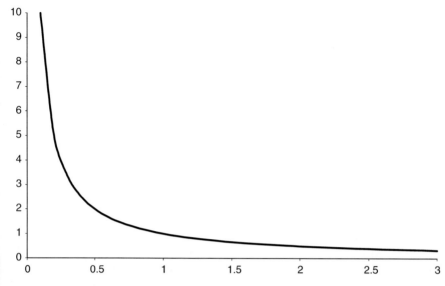

FIGURE 6.9 *An inverse (y = 1/x) curve*

it will not affect our answers. We can add a constant, and as long as we add it to every value of the variable, it will not affect the substantive interpretation of our results. The resulting solution that the regression equation gives us will differ only in the constant, not in the parameter estimates nor in the probability values associated with each of the parameter estimates. Similarly, as we saw previously, we can multiply a variable by any value, and we would find that the unstandardised parameter estimates would alter (in an entirely predictable way) but that, again, the standardised estimates, the probability values, and hence any conclusions that we would draw, would be unaltered. As we saw in Chapter 1, the units in much psychological research are measured on arbitrary scales (we could have used words, paragraphs, pages, sections, chapters or books as measures of how much had been read) so moving from one arbitrary scale to another will not affect our results.

However, when we carry out a non-linear transformation, the range of values of the independent variable will affect the degree to which the transformation causes the line to curve. Figure 6.10 shows the effect of carrying out a quadratic transformation of the values from 0 to 5 (graph A) and 10 to 15 (graph B). The effect of the transformation of the variable when the range is 1 to 5 is much greater than the effect of the transformation when the range is 10 to 15.

You should be aware that some transformations are not possible on some ranges of numbers. There is no log of the number 0, for example, and negative numbers do not have square roots. You will sometimes find that you are required to add a constant to a variable to ensure that the range of numbers is appropriate.

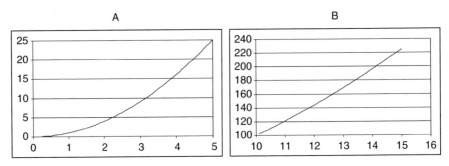

FIGURE 6.10 *The effect of a quadratic transformation for ranges of x from 0 to 5 (A) and 10 to 15 (B)*

6.1.3.2 Transformations and distributions

Transformations have an unfortunate effect: as well as altering the shape of a curve, they also affect the shape of the distribution of a variable. We saw this in Chapter 4 when we tried to make our data closer to a normal distribution. This means that a variable that is normally distributed, or almost normally distributed before transformation, can become non-normal after the transformation. This is particularly problematic in the case of outliers. A case that is close to being an outlier when the data are not transformed may become an extreme case when the data are transformed.

There is, unfortunately, no easy solution to this problem, and any solution that you choose will necessarily prove to be a compromise. One option is to delete the problematic case from the analysis entirely (see Chapter 4 for a fuller discussion of this option). A second option is to delete the problematic value and use the problematic case in the rest of the analysis. We saw in the previous section that adding (or subtracting) a constant to a variable may alter the effect of some transformations. You can investigate whether changing the constant will cause the transformation to have a lesser or greater effect on the outlier.

6.1.3.3 Add hierarchically

An important principle in both science generally and regression specifically is that simpler models are better. If two models are equally good at explaining the observed phenomena, then the simpler one is better. What this means in terms of regression is that if a linear model and a non-linear model describe the data equally well, the linear model is to be preferred – it has fewer parameters. The process of adding the non-linear terms hierarchically, and examining the improvement in R^2, tests whether the non-linear model is an improvement upon the linear model. We said in Chapter 2 never (or hardly ever) to use stepwise regression, and that it is usually better to use hierarchical regression. When adding non-linear terms to a model, never, ever, use stepwise regression. A stepwise regression is likely to add non-linear terms before the linear terms have been added, and to remove the linear term in favour of the non-linear term, making your final model unnecessarily complex and difficult to interpret.

6.1.4 An example of non-linear regression

Table 6.1 shows a dataset consisting of two variables, hassles and anxiety. Hassles is the number of 'daily hassles' that respondents report feeling in the previous two weeks. Daily hassles are small problems that individually may not mean anything, but can build up to cause a large amount of stress and anxiety. Examples of daily hassles, taken from Kohn and MacDonald (1991), include 'Disliking your daily activities', 'Financial

TABLE 6.1 *Dataset 6.1*

hassles	anxiety
38	10
10	12
60	21
90	16
88	27
96	30
1	9
41	7
86	32
59	11
25	13
5	5
3	20
16	9
22	11
41	23
29	6
72	21
55	6
36	19
96	17
36	13
91	46
47	17
63	16
64	23
98	40
81	30
99	43
90	23
95	39
5	16
71	13
82	31
97	37
47	16
30	8
75	16
35	11
78	23

burdens' and 'Unwanted interruptions at work.' anxiety refers to a person's score on a measure of current anxiety.

6.1.4.1 Step 1: fitting the linear model

When we analyse the data using a linear regression model we find that the hassles measure has a large effect on anxiety ($R^2 = 0.533$, $F(1,38) = 42$, $p < 0.05$). The coefficients from the analysis are shown in Table 6.2, showing that anxiety = 0.253 × hassles + 5.423.

TABLE 6.2

	Slope (*b*)	Std error of slope	Standardised slope (beta)	*t*	Sig.
Constant	5.423	2.465		2.199	0.034
hassles	0.253	0.038	0.730	6.592	<0.001

We have a number of reasons to believe that this solution might not be the end of the story. First, and probably most importantly, a linear relationship between these variables is likely not to be based on psychological theory. It seems likely that the importance of the first minor hassle and the 50th minor hassle may not be the same – people often feel that 'just one more thing' might put them 'over the edge', the straw that broke the metaphorical camel's back. However, if we fit a linear model, we assume that every additional hassle has an equal effect.

Second, if we examine the scattergraph that is shown in Figure 6.11 with the line of best fit drawn on it, it seems that the points do not fall along that

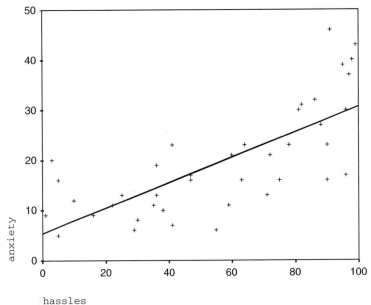

FIGURE 6.11 *Graph of* anxiety *and* hassles

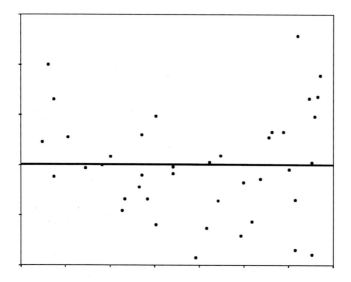

FIGURE 6.12 *Scatterplot of predicted values against residuals*

straight line. The points seem to fall above the line at the upper and lower ends, and below the line in the central region.

Finally, we can examine the plot of residuals (we examined these plots in some detail in Chapter 4) to detect potential non-linearity. Here we plot the predicted values of the dependent variable against the residuals. Figure 6.12 shows this graph and the way in which the residuals form a U-shape. Generally, this pattern (or an inverted-U) suggests that the association is non-linear (Fox, 1991).

Because of the evidence from the various plots and how we think the variables should be related, we will examine the possibility of a quadratic and a cubic relationship between hassle and anxiety.

6.1.4.2 Step 2: creating non-linear terms

The second step in analysing the non-linear relationship is to create the non-linear terms, by calculating the square and the cube of the original hassles score. These non-linear terms are then entered into the calculation hierarchically and the increase in R^2 is calculated and tested for significance. Most software packages will do this automatically, but if they do not, the equation is given in Chapter 2.

The results of this calculation are given in Table 6.3 and Table 6.4. Table 6.3 shows the value of R, R^2 and adjusted R^2 for each of the hierarchy of three models (linear, quadratic and cubic respectively). The table shows that there is a large increase in R, R^2 and adjusted R^2 from model 1 (hassles only) to model 2 (hassles and hassles^2). When hassles^3 is also added to the model, the increase in R and R^2 is very small, and the value of adjusted R^2 actually decreases.

TABLE 6.3 R^2 for three models

IVs	R	R^2	Adjusted R^2
hassles	0.730	0.533	0.521
hassles + hassles²	0.806	0.650	0.631
hassles + hassles² + hassles³	0.808	0.652	0.623

Examining Table 6.4, which shows the significance of the change in R^2, we can see that adding hassles² to the model leads to a significant improvement in R^2, whereas the improvement brought about by adding hassles³ is not significant. (In Table 6.4 the first change is the change in R^2 from the model with a constant only.)

TABLE 6.4 Change in R^2 and significance of change

	\multicolumn{5}{c}{Change statistics}				
	R^2 change	F change	df_1	df_2	Sig. F change
Linear	0.533	43.455	1	38	0.000
Quadratic	0.116	12.267	1	37	0.001
Cubic	0.003	0.263	1	36	0.611

We will therefore accept that the quadratic equation is the most appropriate, and can examine the parameters of the model.

6.1.4.3 Step 3: examining the model parameters

Table 6.5 shows the estimated parameters of the model (labelled slope in the table). Using this information, we can write out the equation. The final equation is:

$$\text{anxiety} = (-0.222) \times \text{hassles} + (0.005) \times \text{hassles}^2 + 13.529$$

TABLE 6.5

	Slope (b)	Std error of slope	Standardised slope (beta)	t	Sig.
(Constant)	13.529	3.169		4.269	0.000
hassles	−0.222	0.140	−0.641	−1.589	0.121
hassles²	0.005	0.001	1.413	3.502	0.001

Some points should be noted with regard to this final equation. First, you will notice that the slope for hassles changed from a positive value to a negative value, and that despite its large standardised value it is not significant. Both of these have occurred because of the collinearity between the hassles variable and the quadratic term hassles² — the correlation between the two is 0.970. The second point is that the unstandardised coefficient for the quadratic term is very small. This small coefficient has occurred because squaring the variable has the effect not only of altering the distribution but also of increasing the variance – because the variance has increased a great deal, any parameter estimates have to shrink to accommodate this.

We now have an equation that we can use to calculate a predicted score on anxiety given any value of hassles. For example, if we want to predict the value of anxiety when hassles = 1, we substitute the value 1 for hassles and we get:

$$\begin{aligned} \text{anxiety} &= (-0.222) \times 1 + (0.005) \times 1^2 + 13.529 \\ &= -0.222 + 0.005 + 13.529 \\ &= 13.312 \end{aligned}$$

Therefore, the value of anxiety when hassles is equal to one is 13.312. We can carry out this calculation for any value of hassles to see how the effect of hassles changes along the line of the curve, for example we could solve the equation when hassles is 10, 20, 30, etc. We may want to know the slope of the line at different values of hassles. This would tell us how much anxiety would increase, if hassles increased by one unit. This might then give us an indication of the level of hassles we would want to stay below. An explanation of the calculation of the slope is a little complex so readers may be advised to skip this section if they feel it is too daunting.

6.1.4.4 Interpreting the quadratic term

We now have an equation that we can use to calculate a predicted anxiety score for any level of hassles. Interpreting the slope coefficient is more difficult, because the slope varies at different levels of hassles, but we can use this equation to calculate the slope coefficient at any level of hassles. If we take our regression equation:

$$\text{anxiety} = 13.529 + \text{hassles} \times (-0.222) + \text{hassles}^2 \times 0.005$$

we can expand this to:

$$\text{anxiety} = 13.529 + \text{hassles} \times (-0.222) + \text{hassles} \times \text{hassles} \times 0.005$$

We can calculate the slope at any value of hassles, by substituting the value of hassles into two of the three times it appears in the equation (the details of why we do this are rather complex, and involve calculus; for more information, see Judd and McClelland, 1989). To calculate the slope at hassles = 1:

$$\text{anxiety} = 1 \times (-0.222) + 1 \times \text{hassles} \times 0.005$$

(Note that the constant is irrelevant to the calculation of the slope, so we have removed it.) We can then simplify this equation to:

$$\text{anxiety} = (-0.222) + \text{hassles} \times 0.005$$

and then to:

$$\text{anxiety} = + \text{hassles} \times 0.005$$

So at the value of hassles = 1, the increase in stress associated with hassles increasing by one is (approximately) 0.005, a very small increase.

We can carry out this calculation for any value of hassles. Some of these values are shown in Table 6.6. It can be seen that as the value of hassles increases, the slope coefficient increases, showing the non-linear nature of the relationship. Although this information is useful, it should always be used in conjunction with a scatterplot including the fitted curve. All standard statistical software should have the ability to draw such plots. The fitted curve is shown in Figure 6.13.

TABLE 6.6 *Slope coefficient at different levels of* hassles

hassles	Slope
10	0.049
20	0.099
30	0.149
40	0.199
50	0.249
60	0.299
70	0.349
80	0.398
90	0.448
100	0.498

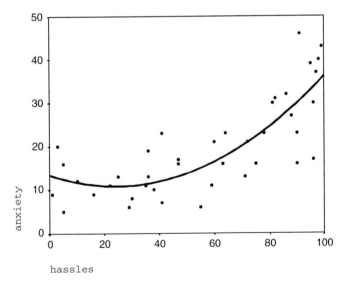

FIGURE 6.13 *Scatterplot showing quadratic line of best fit*

6.2 Logistic regression

6.2.1 The case of the dichotomous dependent variable

We have considered both continuous independent variables (Chapters 1 and 2) and categorical independent variables (Chapter 3) in relation to regression analysis, but in both of these cases we have considered only continuous dependent variables. It is often the case in the social sciences that we wish to examine a categorical dependent variable. We will examine how the standard regression techniques can be adapted to cope with a dichotomous dependent variable, one which can take on one of two values – often 'Yes' and 'No'. If we assign the value 1 to Yes, and 0 to No and then attempt to carry out a regression analysis we are immediately going to encounter several problems. We will demonstrate this in the example below.

Employers are interested in determining whether or not new employees will be successful in their work roles before spending a large amount of time and money training them. A bus company that employs and trains bus drivers is interested in whether they will pass the appropriate driving test. If the bus driver fails the test, even by a tiny margin, they are no use, and similarly whilst passing the test with flying colours is good, it is not really any better than just passing, from the perspective of the company.

Table 6.7 shows a table containing data from an organisation that wanted to assess the efficiency of its employee selection procedures. The first variable, score, is an applicant's score on a test of aptitude towards the job. The second variable, experience, is the number of months of relevant prior experience that the applicant has had before this job. The third variable, pass, is whether the applicant actually passed the test after their training period (1 indicates Yes, 0 indicates No).

To begin with, we will exclude experience from the analysis and focus on score as the independent variable and pass as the dependent variable. If we carry out a standard regression analysis, we find a significant prediction of the dependent variable ($R^2 = 0.097$, $F(1,48) = 5.1$, $p < 0.028$). The slopes and associated test statistics are shown in Table 6.8.

At first glance, this result seems to be a sensible (and statistically significant) solution. It indicates that higher values of score lead to higher values of pass. However, when we look more closely, some major problems in terms of interpretation are revealed.

The first problem occurs when we consider what a slope of 0.110 actually *means* in terms of an individual's performance. If an individual has a value on score that is larger by one than another person's, it does not mean anything to say that the expected value of the outcome of the test will be 0.110 higher. The expected value of the outcome of the test can only be zero or one – the employee can pass the test or fail the test. It is also possible that an individual may do sufficiently well on the aptitude test that we would predict their pass would achieve a score of greater than one. This means that they would do better than passing, but this is not possible, as passing is the only alternative to failing.

APPLYING REGRESSION AND CORRELATION

TABLE 6.7 Dataset 6.2

score	experience	pass
5	24	1
3	15	0
2	12	1
2	24	0
4	18	0
3	9	0
5	8	1
2	12	0
3	24	1
3	6	0
2	15	0
3	20	1
4	8	1
3	15	1
3	18	1
2	16	1
5	5	1
5	18	1
2	8	1
2	20	1
4	24	1
1	18	0
5	12	1
5	6	1
5	6	0
1	15	0
1	12	0
4	6	0
1	15	1
1	6	0
4	16	1
1	10	1
3	12	0
4	26	1
5	2	1
1	12	0
3	18	0
3	3	0
1	24	1
2	8	0
1	9	0
4	18	0
4	22	1
5	3	1
4	12	0
4	24	1
2	18	1
2	6	0
1	8	0
5	12	0

TABLE 6.8

	Slope (b)	Std error of slope	Standardised slope (beta)	t	Sig.
(Constant)	0.190	0.161		1.182	0.243
score	0.110	0.048	0.311	2.270	0.028

A second problem emerges when we examine the residual plots. First, if we look at the probability plot (see Chapter 4) of the residuals, which is shown in Figure 6.14, they are not falling along the diagonal, as we would expect if they were normally distributed. This means that these data have violated the assumption of normal distribution, and we therefore cannot trust either the parameter estimates or the standard errors (and therefore significance tests).

Second, if we examine the scatterplot of predicted values against the residuals (Figure 6.15), we do not find the pattern that we would expect.

We need some way instead to transform the data so that we can proceed with the analysis. The transformation is a little different from the transformations that we encountered earlier in this chapter. In non-linear regression, some sort of non-linear transformation changes the independent variables. When we have a dichotomous dependent variable, we no longer transform the independent variables; rather we transform the dependent variable. The transformation required in this situation is called the logit transformation. When the logit transformation is used in regression analysis, it is called logistic regression.

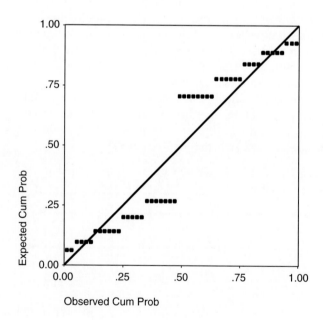

FIGURE 6.14 *Normal probability plot of residuals from linear regression with a dichotomous dependent variable*

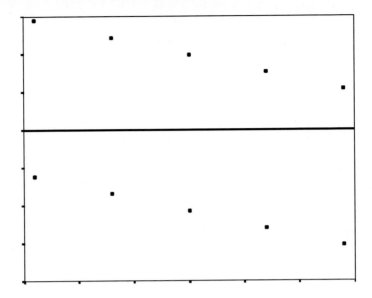

FIGURE 6.15 *Plots of predicted values against residuals*

6.2.2 The logit transformation

When the dependent variable is expressed simply as pass or fail, it is difficult to make a prediction about how somebody might do. In 'real life' we would say that they are likely to fail, they may pass, or would almost definitely pass. In logistic regression we want to do the same thing, so rather than talk of pass and fail, we can convert the data into probabilities, as shown in Table 6.9. This shows the number of people (N) who passed or failed the exam broken down by score. From these the probability of a person of each score group passing or failing the exam is calculated (P). We can now talk in terms of values between zero and one, because we are now talking about probabilities. We can say (in this sample) that the probability of a person who scores 1 on the aptitude measure passing the final test is 0.3.

TABLE 6.9 *Number of employees passing the test, broken down by score, with probabilities*

	Score	1	2	3	4	5
Pass	N	7	5	6	4	2
	P	0.7	0.5	0.6	0.4	0.2
Fail	N	3	5	4	6	8
	P	0.3	0.5	0.4	0.6	0.8

However, if we use these data we will still face the problem that the predicted values can extend below zero and above one. When values represent probabilities, it is not possible for a predicted value to be less than zero or greater than one. One way around this limitation is to convert the probability into odds. Odds are more familiar to most people than probabilities – and they probably were to you, before you started to study statistics. Even people who do not gamble are familiar with the concept of odds, and tend to think in odds, rather than probabilities. If we were to ask you what the chances of throwing a six on a fair, six-sided die were, you would probably say '1 in 6', not '0.16667', and similarly, if we asked you the chances of not throwing a six, you would probably say '5 in 6', not '0.83333'. However, we want the odds to be represented as a single number (we cannot put '1 in 6' into an equation), and this is called the odds ratio.

The odds ratio of an event happening is calculated using the formula:

$$\text{Odds ratio} = \frac{P(\text{event})}{1 - P(\text{event})}$$

where $P(\text{event})$ refers to the probability of a particular event occurring, and $1 - P(\text{event})$ refers to the probability of the event not occurring.

The odds ratio for not throwing a six on a die is equal to:

$$\text{Odds ratio} = \frac{0.833}{0.167}$$
$$= 5$$

So, we are five times more likely not to throw a six (five times out of six) than we are to throw a six (one time in six).

Similarly, we can convert from an odds ratio to a probability by using the formula:

$$P(\text{event}) = \text{odds}(\text{event})/[1 + \text{odds}(\text{event})]$$

If the odds ratio of not throwing a six is equal to five:

$$P = 5/[1 + 5]$$
$$= 5/6$$
$$= 0.833$$

Now we can extend Table 6.9 to include odds, as shown in Table 6.10. This table shows that a person who achieves a score of 1 on the aptitude measure is 2.33 times more likely to fail the test than to pass it. A person who scores 2 is equally likely to pass as to fail, and so on.

Odds ratios have one very large advantage over probabilities. We now have a measure that can extend above one – there is no danger of predicting that a person will have a value that is too high to exist. An odds ratio of 2, or of 50, or of 1000 is OK, and is not out of bounds.

TABLE 6.10 *Number of employees passing and failing, with probabilities and odds*

	Score	1	2	3	4	5
Fail	N	7	5	6	4	2
	P	0.7	0.5	0.6	0.4	0.2
Pass	N	3	5	4	6	8
	P	0.3	0.5	0.4	0.6	0.8
	Odds	2.33	1.00	1.50	0.67	0.25

However, we still cannot extend the value below zero, so we need to do something different. An odds ratio of 0 represents an event that will never happen, and is equivalent to a probability of 0. So we need to do one more thing to allow the predicted value to go below zero. This final stage is to take the natural logarithm (log) of the odds ratio, which gives the logit. The natural log of a value is usually found using the *log* function in mathematical or statistical computer programs, or *ln* button on a scientific calculator. The natural log of the odds ratio can stretch to plus or minus infinity ($\pm\infty$), but when we convert back to a probability, it is still bounded between zero and one. Table 6.11 shows the values expressed in their raw form, as probabilities, odds and logits. It should be noted that the probability, the odds and the logit are all ways of expressing the same thing — if you know the probability, you can work out the odds and the logit. If you know the logit, you can work out the odds and the probability by reversing the process.

TABLE 6.11

	Score	1	2	3	4	5
Fail	N	7	5	6	4	2
	P	0.7	0.5	0.6	0.4	0.2
Pass	N	3	5	4	6	8
	P	0.3	0.5	0.4	0.6	0.8
	Odds	2.33	1.00	1.50	0.67	0.25
	Logit	0.37	0.00	0.18	−0.18	−0.60

Figure 6.16 shows the logit (*x*-axis) plotted against the probability (*y*-axis). The probabilities can get closer and closer to zero and one, but they can never reach them.

6.2.3 Using the logit: logistic regression

In linear regression, the calculations minimise the sum of the squared residuals to estimate the parameters. This is known as ordinary least squares (OLS) regression. The equations that are used to estimate the parameters are known as closed form equations (see Appendix 1): you (or more likely the

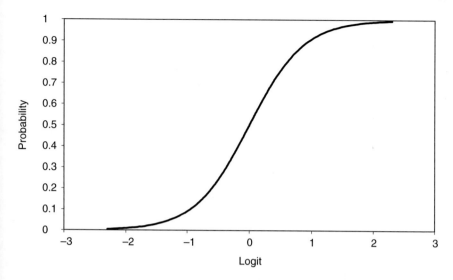

FIGURE 6.16 *The logit curve*

computer) work through the equations and obtain an answer. The equations in logistic regression are not of this closed form type and so OLS estimation cannot be used. Instead, we use an estimator called maximum likelihood (ML). We will give a very brief overview of the way ML works, and because we have probably oversimplified to the point of being incorrect, we direct the interested reader to Eliason (1993).

ML estimation works iteratively – it has a guess at what the parameter estimates are, and then calculates a measure of how well the parameter estimates fit the data. This measure is called the log likelihood function, and the larger it is the better the model fits the data. The parameter estimates are then tweaked a little, and the log likelihood function is re-estimated. If it shows an improvement, the estimates are kept, but if it shows deterioration, the estimates are discarded and the model is tweaked a different way. When the model cannot be improved by tweaking in any way, it is said to have converged, and the parameter estimates are retained. In this way, the ML function attempts to maximise the likelihood of obtaining the observed values of the dependent variable, given the independent variables. (We mentioned in Chapter 1 that the mean was a special case of a least squares estimate. OLS is also a special case of an ML estimate – it would be possible to use ML to estimate linear regression parameters, and therefore the mean of a sample. If we did use ML rather than OLS we would get exactly the same answer. However, in much the same way that we use the mean because it is simpler than calculating the OLS estimate, and gives the same result, we use the OLS estimate, because it is simpler than using ML, and gives the same result.)

The output from a logistic regression analysis looks a little different from the output of an OLS regression analysis but is similar in many ways.

6.2.4 An annotated example of logistic regression

In this section, we will examine the output from a logistic regression model using SPSS. Other statistical packages produce very similar output. The first part of the output presents the initial statistics — this enables us to evaluate the overall model, in much the same way that multiple R, R^2, adjusted R^2 and ANOVA enable us to evaluate an OLS regression.

This part of the output provides us with an initial log likelihood function. In the previous section we said that the ML estimation procedure attempts to maximise the value of the log likelihood function in much the same way that OLS regression attempts to maximise the value of R^2 (by minimising the sum of squared residuals). The log likelihood function is multiplied by -2, and referred to as the -2 log likelihood (-2LL). This conversion has two advantages. First, the log likelihood is a negative number, so if it is multiplied by -2 it becomes positive (and therefore slightly easier to deal with). Second, by multiplying it by 2 it also becomes distributed approximately as χ^2, enabling us to assess its significance. Because we have now converted it to a positive value, lower values of -2LL indicate a better fit. When using OLS regression we know that the minimum value for multiple R is zero, and we will get this value if we have accounted for none of the variance in the dependent variable. Similarly, if we have perfect prediction multiple, R will be one. If we have perfect prediction in a logistic regression we know that -2LL will be zero, but we do not know the maximum value it can attain. We need to know this figure to see if our value is an improvement (in much the same way that we need to test multiple R against zero). To give us this value, the initial -2LL is given, which is the value when there are no dependent variables in the model (only a constant). This value provides a baseline against which we can compare subsequent models. This baseline model is equivalent to the linear regression model in which we only had the mean, and none of the independent variables had been entered.

```
Dependent Variable.. PASS
Beginning Block Number 0. Initial Log Likelihood Function
-2 Log Likelihood 69.234697
* Constant is included in the model.
```

The next lines tell us that the iterations converged on a solution. Very occasionally, you may find that the estimates do not converge within the maximum number of iterations allowed. Usually you will find that simply increasing the number of iterations will solve this problem. In the days when computers were much slower, and one analysis would take hours, this was more of an issue than it is now.

```
Estimation terminated at iteration number 3 because
parameter estimates changed by less than .001
```

Following this, the output shows -2LL for the model. This is equivalent to the value of multiple R, or F in the ANOVA section, as it is used to assess the overall significance of the model. Included with the value of -2LL for the model are several other measures designed to assess the fit of the model. There are two alternative values for R^2 given in the SPSS output – there is no exact analogy for R^2 in logistic regression. For a fuller discussion of the methods of assessing fit in logistic regression, see Menard (1995).

```
-2 Log Likelihood    64.245
Goodness of Fit      49.795
Cox & Snell  - R^2     .095
Nagelkerke   - R^2     .127
```

Finally in this section are the χ^2-value for the model, the block and the step. In this case they are all the same. The model is the overall χ^2 for all of the variables entered into the equation. We want this to be significant (usually at the 0.05 level). If it is statistically significant, we can say that it is unlikely that chance effects alone would predict the dependent variable as well as our model. The block would differ (and we would be interested in it) if we were carrying out a hierarchical logistic regression as it would give the χ^2-value for the variables that we had entered in that block (we will see this in the next section). The step would differ if we were carrying out a stepwise logistic regression (and we do not need to repeat what we said in Chapter 2 regarding stepwise regression). χ^2 is used to assess the significance of the model in much the same way as F in ANOVA is used to assess the significance of the model. χ^2 is calculated using:

$$\chi^2 = (\text{-2LL}_0) - (\text{2LL}_M)$$

where -2LL$_0$ is value of -2LL for the baseline, or null, model, that is the model before the variable(s) were entered, and -2LL$_M$ is the value of -2LL for the model after the variable(s) were entered. For this example χ^2 is given as:

$$69.23 - 64.25 = 4.98$$

This is distributed as χ^2 with df equal to the number of independent variables in the null model minus the number of independent values in the model under test. There were no independent variables in the null model, and one in the model under test, so $df = 1$.

	Chi-Square	df	Significance
Model	4.990	1	.0255
Block	4.990	1	.0255
Step	4.990	1	.0255

In OLS regression, we could compare the predicted values for the dependent variable with the actual values of the dependent variable to see how well our model was able to predict the dependent variable. You may recall that the correlation between the dependent variable and the predicted values of the dependent variable is equal to multiple R. In logistic regression, we do not want to know if the predicted values are *close* to the real values – the variable can only take on two possible values after all. We want to know if the predicted values are *equal* to the real values of the dependent variables.

The next section provides us with information about predicted and observed (real) values of the dependent variable, in the form of a 2 × 2 classification table.

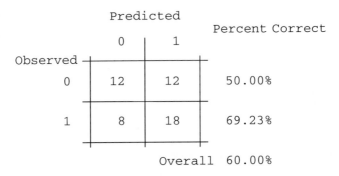

```
Classification Table for PASS
The Cut Value is .50
```

	Predicted 0	Predicted 1	Percent Correct
Observed 0	12	12	50.00%
Observed 1	8	18	69.23%
		Overall	60.00%

The two rows provide the actual values of the dependent variable, and the two columns the predicted values of the dependent variable, based on the model. Here, 24 employees failed the test. Using a model we would have predicted that 12 of them would pass and 12 of them would fail, a success rate of 50%. Of the 24 employees who passed the test, we would have predicted that 8 of them would have failed and 18 of them would have passed, a success rate of 69.23%. Overall, according to the table, we would have a success rate of 60%, which is better than chance (50%), but not much.

Finally, we get to the actual parameter estimates in the equation. The value of the slope coefficient is called B, and is similar to the value of the slope in OLS regression: it is the amount of change in the dependent variable associated with a change of one unit in the independent variable. However, recall that this is a change in the log odds ratio, rather than an absolute change.

```
- - - - - - - - - - Variables in the Equation - - - - - - - - - -
```

Variable	B	S.E.	Wald	df	Sig	R	Exp(B)
SCORE	.4674	.2188	4.5658	1	.0326	.1925	1.5959
Constant	-1.3137	.7135	3.3898	1	.0656		

To calculate what this means in terms of the predicted outcome, we must use the equation:

$$\text{logit(pass)} = B \times \text{score} + \text{constant}$$

We then need to work backwards to convert the logit to an odds ratio, using:

$$\text{odds ratio} = \exp(\text{logit(pass)})$$

and finally to a probability, using:

$$p = \text{odds}/(1 + \text{odds})$$

Thus for a person who scores 1 on the score variable:

$$\text{logit(pass)} = 0.47 \times 1 - 1.31 = -0.84$$
$$\text{odds} = \exp(0.84) = 0.43$$
$$p = 0.43/(0.43 + 1) = 0.30$$

and for a person who scores 5:

$$\text{logit(test)} = 0.47 \times 5 - 1.31 = 1.04$$
$$\text{odds} = \exp(1.04) = 2.82$$
$$p = 2.82/(2.82 + 1) = 0.74$$

So a person who scores 1 on the aptitude measure has a probability of 0.30 of passing the test, and a person who scores 5 on the aptitude measure has a probability of 0.74 of passing the test. (You may want to work through the probability of someone passing the test who scores a ridiculously high score on the aptitude measure, to demonstrate that it is not possible to have a probability of passing of greater than one.)

A different way of interpreting B is to consider the exponential of the score as being a multiplier of the odds ratio, given an increase of one unit in the independent variable. The exponential of the score is provided in the final column of the output to allow us to do this. If we take the exponential of the constant, we find the odds ratio for an individual who has scored zero on the aptitude measure:

$$\text{odds} = \exp(-1.31) = 0.27$$

We can convert this into a probability in the usual way:

$$p = 0.27/(0.27 + 1) = 0.21$$

An individual who scores one unit higher (i.e. scores 1) on the aptitude measure can multiply their odds ratio by $\exp(B)$, which in this case is 1.6. Consequently, this individual would have an odds ratio of:

$$\text{odds} = 0.27 \times 1.6 = 0.43$$

You might notice, before we bother to calculate it, that this is the same odds ratio that we calculated in the other way. Thus, a B parameter can be interpreted as a multiplier for an odds ratio, which means that we can interpret B in the presence of other independent variables in much the same way that we can in multiple linear regression.

The column after B is the standard error of B, which is used to calculate confidence intervals around parameter estimates and predictions in much the same way as the confidence interval is used in OLS regression. The Wald test is used in the same way as the t-test in OLS regression, that is to calculate the significance of the parameter. The Wald test is distributed as χ^2, and its associated df and p-value are shown in the table. As in OLS regression, only variables with a statistically significant B should be interpreted.

For completeness R is a similar statistic to the standardised slope in OLS regression — it gives a standardised measure of the effect of the variable.

6.2.5 Hierarchical logistic regression

In the same way as variables may be entered and assessed hierarchically in multiple OLS regression, variables may be entered and assessed in multiple logistic regression. In this example, we show and describe the output from SPSS when a second variable in the analysis has been added. The calculations to assess the improvement in the model are carried out automatically by SPSS (they are not complicated to carry out if they are not given by the statistics package that you are using).

The company in our hypothetical example may also have assessed the amount of relevant previous experience that potential employees have, and be interested to know whether this will improve their classifications. To find out we enter the experience variable hierarchically and assess whether the variable has a significant effect in improving the prediction of performance in test beyond what we already knew from the score variable.

The first part of the output simply repeats the output that we saw in the previous section, so we do not reproduce it here. The output starts with an introductory section, which tells us that it is on block number 2 and gives the overall measures of model fit that we saw in the previous section.

```
Beginning Block Number 2. Method: Enter

Variable(s) Entered on Step Number
1..        EXP

Estimation terminated at iteration number 3 because
Log Likelihood decreased by less than .01 percent.

-2 Log Likelihood            59.066
Goodness of Fit              48.343
Cox & Snell - R^2              .184
Nagelkerke - R^2               .245
```

	Chi-Square	df	Significance
Model	10.169	2	.0062
Block	5.178	1	.0229
Step	5.178	1	.0229

First, we have the model χ^2. This compares the model that has two independent variables with the model that has no independent variables, and is significant. The second χ^2 given is the block χ^2 which compares the value of -2LL of the model that has the independent experience and score variables in it with the value of -2LL of the model that contains only score. Because this result is significant, we can conclude that the improvements in the model due to the addition of experience have not arisen by chance.

The next section reproduces the classification table. It can be seen that the classification has improved. The inclusion of the experience variable has increased the percentage of correct classifications from 60% to 72%.

Classification Table for PASS
The Cut Value is .50

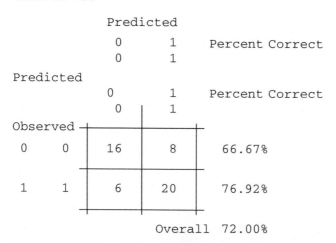

		Predicted		Percent Correct
		0	1	
Observed	0	16	8	66.67%
	1	6	20	76.92%
			Overall	72.00%

Finally, the effects, standard errors and other test statistics are reported.

--------- Variables in the Equation ----------

Variable	B	S.E.	Wald	df	Sig	R
SCORE	.5486	.2345	5.4727	1	.0193	.2325
EXP	.1115	.0521	4.5771	1	.0324	.2003
Constant	-3.0495	1.1456	7.0860	1	.0078	

The table shows the values for B, and we can conclude that the effects of both variables are significant (using the Wald test).

6.2.6 Polynomial logistic regression

In this chapter, we have only considered the case where the dependent variable is a categorical variable with two possible values. The formal name for this type of analysis is binomial logistic regression. It is possible to carry out polynomial (or multinomial) logistic regression in the situation where the dependent variable has more than two categories, although this is implemented in fewer statistical packages. The interested reader is advised to read the book by Hutcheson and Sofreniou (1999).

Further reading

For non-linear regression, see Pedhazur (1997) *Multiple regression in behavioral research: explanation and prediction*, Chapter 13.

For logistic regression see Hutcheson and Sofreniou (1999) *The multivariate social scientist: introductory statistics using generalised linear models*, or Menard (1995) *Applied logistic regression analysis*.

7 Moderator and mediator analysis

7.1 Introduction

In Chapter 1 we considered a simple bivariate relationship between a single predictor variable and a single outcome variable (x and y). In Chapter 2 we extended this basic model to include two or more predictor variables (x_1, x_2). Having two or more predictor variables allowed us to examine the effect of one predictor variable while controlling for the other predictor variable(s). In this chapter we will consider ways in which predictor variables may 'work together' to affect an outcome variable. The first way that two variables may work together is as a moderator effect. A variable is said to moderate the effect of a second variable if the effect of the second variable depends upon the level of the first variable. Readers familiar with ANOVA may recognise this as an interaction effect. It will not surprise readers who have approached this book sequentially to discover that interactions in ANOVA are a special case of a regression analysis. The second way that two variables may work together is as a mediated relationship. A mediated relationship is said to occur if a predictor variable has its effect on the outcome variable via a second predictor variable. We hinted at the mediator approach when we looked at hierarchical regression in Chapter 2.

7.2 Moderator analysis

You may be familiar with the notion of moderator analysis from an ANOVA framework, where the analysis is usually known as an interaction. The different terms moderator effect and interaction effect are fairly interchangeable. Moderator effects are usually used to refer to cases where at least one of the independent variables is continuous. Interaction effects in ANOVA are a special case of a moderator with categorical variables (recall from Chapter 2 that ANOVA is a special case of a regression analysis).

Moderated relationships are very simple to think about in real-life terms. In statistical terms, however, they can be quite bewildering. A third variable (Z) is said to moderate the relationship between two other variables (X and Y) if the degree of relationship between X and Y is affected by the level of Z. As we said in Chapter 4, G.H. McClelland (personal communication) has used the phrase 'different slopes for different folks' to describe this kind of relationship. Some examples from everyday life should make things clearer. (We will risk being

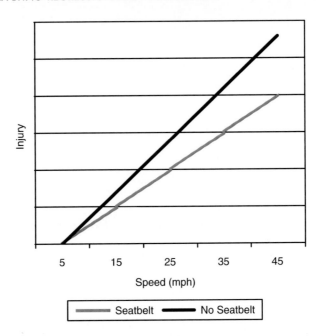

FIGURE 7.1 *Lines of best fit for people wearing and not wearing a seatbelt*

seen as facetious when we give these examples, but we feel that students have sufficient difficulty with the concept to justify risking this charge.)

If you drive your car into a wall, the speed at which the car is travelling will affect how much you are hurt. Generally, the higher the speed of the car (X) the more you will be hurt (Y). This relationship is moderated by whether or not you are wearing a seatbelt. If you are travelling at 5 mph (8 kph) when you hit the wall you probably will not hurt yourself. If you are travelling at 30 mph (48 kph) you are likely to hurt yourself at least a little, even if you are wearing a seatbelt. However, if you are not wearing a seatbelt you will probably hurt yourself more seriously. Such a hypothetical relationship is shown in Figure 7.1. Thus the wearing of a seatbelt moderates the relationship between speed and injury.

In the previous example we have a relationship between two variables, one a continuous predictor variable (`speed`) and the other a continuous outcome variable (`injury`), and these two are moderated by a categorical variable (`seatbelt`).

It is also possible to find that a continuous variable moderates the relationship between two other continuous variables. Sailing is a relatively expensive hobby. People who are better-off are more likely to go sailing than people who are not so well-off, but there is not a clear relationship between the amount of money that people have and the amount of time that they spend going sailing. A third variable, how much people like going sailing, moderates the relationship between the two variables. The relationship

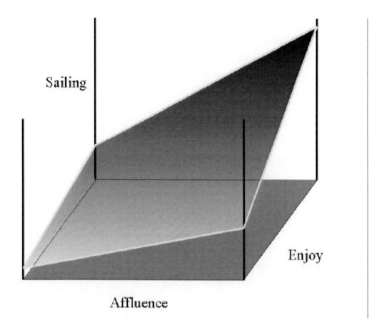

FIGURE 7.2 *A three-dimensional graph of affluence, enjoyment of sailing and time spent sailing*

between these three variables becomes difficult to represent graphically, because this requires a graph with three dimensions, and we are limited to two-dimensional paper. However, we will not allow this limitation to stop us from trying.

Figure 7.2 shows a three-dimensional graph of the amount of money people earn (affluence, on the *x*-axis), the amount of time they spend going sailing (sailing, on the *y*-axis) and the amount they enjoy sailing (enjoy, on the *z*-axis). It can be seen in the graph that the more people enjoy sailing, the more they go sailing, but this does not tell the whole story. The slope is not consistent at all levels of affluence. At the far left-hand side of the graph, people with less money go sailing a little more if they enjoy sailing more. On the right-hand side of the graph, where we have the people who earn more, those more affluent who do not enjoy it go sailing only a little more than the people who are less affluent, but those more affluent people who do enjoy sailing spend a great deal of time sailing. We can see that the slope is different at each end of the *x*-axis (different slopes for different folks) and we conclude that the enjoyment of sailing moderates the relationship between affluence and sailing.

In every case of regression where we wish to test for a moderated relationship, we need to create a new variable to represent the moderator effect. We will begin by describing a regression analysis with two categorical predictor variables plus a moderator term. Next, we will examine the case of a regression analysis with one continuous predictor variable moderated by one categorical predictor variable. Finally we will examine a case with two continuous variables interacting. This book (because of space limitation) provides only a brief guide to the analysis of moderator effects in regression analysis – we do not explain any alternative approaches, or expand on the reasoning to any great extent. However, there are available two excellent texts on this type of analysis: Aiken and West (1991) and Jaccard, Turrisi and Wan (1990).

7.2.1 Two categorical variables

One of the most common designs in experimental psychology is the 2 × 2 factorial ANOVA. In this experimental design there are two independent variables (or in ANOVA terms, factors) and one dependent variable. The experimenter assesses the effect of each predictor variable on the dependent variable and then assesses the interaction effect of the two predictor variables on the dependent variable. So instead of only two effects to be estimated (the effects of the predictor variables), there are now three (the effects of the predictor variables plus the effect of the interaction).

The classic example is the study by Godden and Baddeley (1975), examining context-dependent memory. In this study divers were asked to learn material while on the ground (dry) or underwater (wet) and are then tested on this material, either underwater (wet), or on the ground (dry). Thus we have two predictor variables (learn and test) each of which can take on one of two values, for each participant. The outcome is the number of correct responses (score). We have four groups of participants (Group 1: learn wet, test wet; Group 2, learn wet, test dry; Group 3, learn dry, test wet; Group 4, learn dry, test dry). This is a typical 2 × 2 independent-groups ANOVA design.

Table 7.1 gives hypothetical data from such an experiment, which contains 10 people in each group.

Table 7.2 shows the mean scores for each of the four groups and the group total means. Figure 7.3 shows a bar chart of these results. The table shows that there is a small effect of learn: the wet learning group (ignoring testing for the moment) have a mean score of 26.00, and scored slightly higher than the dry group (mean of 23.75). There is a slightly larger effect of testing environment, dry (mean = 26.80) being superior to wet (mean = 22.95).

The interesting aspect of the results in this example is the interaction between the two variables.

We saw from the table that being tested in the dry seemed to lead to better performance than being tested in the wet. But a glance at the chart shown in Figure 7.3 shows that we need to modify that statement: being tested in the dry does lead to better performance than being tested in the wet *but only if the person learned in the dry*. If a person learned in the wet, they would be better-off being tested in the wet environment. We could describe this effect in a number of different, but equivalent, ways. We could say that there is an interaction effect of testing and learning environment in predicting score.

TABLE 7.1 Dataset 7.1

learn	test	score
Dry	Dry	29
Dry	Dry	29
Dry	Dry	31
Dry	Dry	33
Dry	Dry	34
Dry	Dry	30
Dry	Dry	28
Dry	Dry	33
Dry	Dry	34
Dry	Dry	27
Wet	Dry	23
Wet	Dry	21
Wet	Dry	20
Wet	Dry	22
Wet	Dry	25
Wet	Dry	17
Wet	Dry	25
Wet	Dry	25
Wet	Dry	27
Wet	Dry	23
Dry	Wet	15
Dry	Wet	20
Dry	Wet	24
Dry	Wet	11
Dry	Wet	19
Dry	Wet	13
Dry	Wet	15
Dry	Wet	17
Dry	Wet	17
Dry	Wet	16
Wet	Wet	24
Wet	Wet	20
Wet	Wet	32
Wet	Wet	31
Wet	Wet	31
Wet	Wet	26
Wet	Wet	29
Wet	Wet	27
Wet	Wet	34
Wet	Wet	38

TABLE 7.2 Means and SDs (in brackets) for data in Table 7.1

		test		Group total
		Dry	Wet	
learn	Dry	30.80	16.70	23.75
		(2.57)	(3.68)	(7.87)
	Wet	22.80	29.20	26.00
		(2.94)	(5.18)	(5.25)
Group total	Mean	26.80	22.95	24.88
		(4.91)	(7.76)	(6.70)

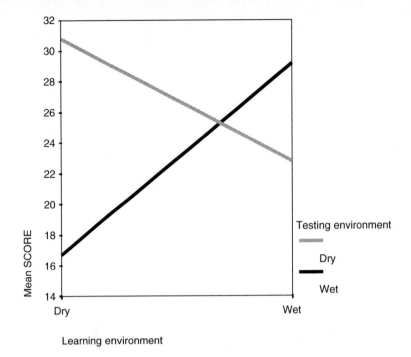

FIGURE 7.3 *Line chart of means from context-dependent memory study*

We could also say that testing environment moderates the effect of learning environment, or we could say that we have 'different slopes for different folks'.

The usual analysis technique applied to these data would be a 2 × 2 ANOVA and the results of such an analysis are presented in Table 7.3.

TABLE 7.3 *Results of ANOVA*

Source	df	F	Sig.
Corrected model	3	29.934	<0.001
learn	1	3.638	0.064
test	1	10.653	0.002
learn × test	1	75.509	<0.001
Error	36		
Total	40		
Corrected total	39		

From the table we can see that the variable learn does not have a significant effect on score, but test does have a significant effect. Below this we see the learn × test effect which represents the influence of the interaction on the outcome variable. This is statistically significant, which indicates that we are not just interested in the individual effects of learn and test, but rather the combined effect of learn and test.

However, we could also carry out this analysis using effect regression. First we recode the `learn` and `test` variables to a coding scheme (see Chapter 3). We code 'dry' in each variable as −1 and wet as 1. We label these two new variables `indlearn` and `indtest` (for indicator-coded learning and testing). The second stage is to create the interaction term; this is done automatically for ANOVA in most computer packages. We must create a new variable to represent the interaction, which is the product of `indlearn` and `indtest`. We have termed this variable `lxt`, and it is calculated using:

$$lxt = indtest \times indlearn$$

Table 7.4 shows how the old variable codes have been altered to create the new variables.

TABLE 7.4

learn	indlearn	test	indtest	lxt
Dry	−1	Dry	−1	1
Dry	−1	Wet	1	−1
Wet	1	Dry	−1	−1
Wet	1	Wet	1	1

If these three variables are entered into a regression analysis as predictor variables we find that we have a significant result for the model ($R = 0.845$, $R^2 = 0.714$, Adj. $R^2 = 0.690$, $F(3,36) = 29.934$, $p < 0.05$). This result shows that we have a large effect and have accounted for a large proportion of the variance in the `score` variables. Interestingly, if we compare these results with the results in Table 7.3, we find that the values of F and df are identical.

The rest of the regression output is shown in Table 7.5. First we can see that the grand mean (24.88 from Table 7.3) is reproduced in Table 7.5 as the constant (the t-value and associated probability are a test that the global total mean > 0, which is rather uninteresting in this case). We also find that the significance values are all equal to those we found in ANOVA.

TABLE 7.5 *Output from analysis using regression*

	Slope (b)	Std error of slope	Standardised slope (beta)	t	Sig.
Constant	24.88	0.59		42.18	<0.001
indlearn	1.12	0.59	0.17	1.91	0.064
indtest	−1.92	0.59	−0.29	−3.26	0.002
lxt	5.12	0.59	0.78	8.69	<0.001

The unstandardised coefficients for the two predictors represent the difference between each group and the total mean. The coefficient for `indlearn` is 1.12; Table 7.2 shows that the grand mean is 24.88. The overall mean for the dry learning group is 23.75, and:

$$\text{dry: } 23.75 = 24.88 + (-1)(1.12)$$

(within rounding error). We use the multiplier −1 because this was the code for the dry group. Similarly for the wet group:

$$\text{wet: } 26.00 = 24.88 + 1(1.12)$$

and we can do the same for `indtest`:

$$\text{dry: } 26.80 = 24.88 + (-1)(-1.92)$$
$$\text{wet: } 22.95 = 24.88 + 1(-1.92)$$

Note that the calculation for the testing environment is slightly complicated by the fact that the coefficient is negative.

Under the null hypothesis (H_0) of no main effects (i.e. no effect of `learn` and `test`), each group has an expected value equal to the grand mean, so each of the parameters for the main effects tells us the extent to which each group mean varies from the grand mean. The null hypothesis (H_0) for the interaction effect is slightly more complex. This hypothesis says that the mean for each of the four groups will be equal to the value predicted by the main effects. If this were true it would mean that the means could vary for each group but the slopes would remain the same.

To calculate the predicted values (y) under the null hypothesis of no interaction effect, we calculate the expected values for each cell using the grand mean and the slope coefficients. The calculation is the standard regression equation:

$$y = b_1 x_1 + b_2 x_2 + c$$

So ignoring the interaction effect:

$$\text{score} = \text{indlearn} \times 1.12 + \text{indtest} \times -1.92 + 24.88$$

The results of this calculation for each table are shown in Table 7.6. The first column shows the groups, the second the values for `indlearn` and `indtest`. Then the calculation under the null hypothesis of no interaction effect is shown. The observed value of the mean for that group is also presented. Finally the difference is shown between the observed value for each cell and the expected value for each cell. These are all equal to 5.12 (or −5.12), which, you will notice, is the value of the parameter for the interaction effect.

TABLE 7.6

Group	indlearn	indtest	H_0 expected value	Observed value	O − E
Dry, dry	−1	−1	$24.88 + (-1)(1.12) + (-1)(-1.92) = 25.68$	30.80	−5.12
Dry, wet	−1	1	$24.88 + (-1)(1.12) + 1(-1.92) = 21.84$	16.70	5.12
Wet, dry	1	−1	$24.88 + 1(1.12) + (-1)(-1.92) = 27.92$	22.80	5.12
Wet, wet	1	1	$24.88 + 1(1.12) + 1(-1.92) = 24.08$	29.20	−5.12

By now, we feel (hope!) we have convinced you that the regression analysis and the ANOVA approach are equivalent. We may have left you asking why we have bothered to introduce the regression approach. The ANOVA approach is surely far simpler to carry out: it does not require the additional steps in data preparation and we can get the information we need from a table of means and a graph. So why are we advocating bothering with this long-winded method?

The answer is that we are not advocating using regression in these circumstances. Given this particular dataset to analyse we would not hesitate to use ANOVA. However, ANOVA is a very restrictive method – it can go little further than we have just gone. Because ANOVA is so limiting, we need regression in case we want to do more complex analysis. In the next sections, using much of the same logic and ideas, we will examine how the analysis that we undertook using two categorical variables can be generalised so that we can use continuous variables. Before we do this we would like to explain a couple more things about the regression approach to ANOVA.

7.2.1.1 Balanced and unbalanced designs
The analysis we have discussed so far has involved an experiment with balanced design – where all of the groups have equal numbers of participants. The balanced design has the advantage that all of the predictor variables are uncorrelated, and so there are no adjustments to be made due to the predictor variables sharing variance. Psychologists are remarkably fond of balanced design even when such a design is not necessarily the most appropriate (McClelland, 1997). Sometimes psychologists use balanced design because they believe (erroneously) that ANOVA cannot be used to analyse unbalanced designs. This has never been the case: it is *much* harder to analyse an unbalanced design by hand – the calculations are far more complex. If a computer is doing the calculations for you, it does not matter if the calculations are more complex.[1]

7.2.1.2 Designs larger than 2 × 2
In this example we have analysed only a 2 × 2 design. However, it is possible to analyse data from studies that have a greater number of categorical predictor variables, and also a greater number of levels for each predictor variable, but the number of interaction terms that must be created increases very rapidly. For a 2 × 2 × 2 design where we have factors A, B and C, we must create three indicator-coded variables along with three two-way interaction terms ($A \times B$, $A \times C$, $B \times C$) and one three-way term ($A \times B \times C$). If we have four variables, with two levels each, we need four indicators for the factors (A, B, C, D), six variables to represent the two-way interactions ($A \times B$, $A \times C$, $A \times D$, $B \times C$, $B \times D$, $C \times D$), three variables to represent the three-way interactions ($A \times B \times C$, $A \times B \times D$, $B \times C \times D$) and one to represent the four-way interaction ($A \times B \times C \times D$).

If we are analysing a 3 × 2 design, we need to represent the first predictor using two indicator variables ($A1$ and $A2$), and the second using one indicator

174 APPLYING REGRESSION AND CORRELATION

(*B*). We then need to create the interaction by using $B \times A1$ and $B \times A2$. We do not have space to explore fully the possible coding mechanisms, but the interested reader is directed to Rutherford (2000) for a full and detailed explanation of these, and many more, coding mechanisms.

7.2.2 Categorical and continuous variables

The procedure we have described above generalises very easily to the situation where we have one categorical independent variable and one continuous independent variable, along with the usual continuous dependent variable. In this case we have a relationship between two continuous variables, which is moderated by a third categorical variable.

Consider a psychologist who is interested in assessing the effects that stressful life events and marital status have on experienced stress levels. The data in Table 7.7 consist of both categorical and continuous data. In the table, events refers to an individual's score on a life events scale, which assesses the number and severity of recent stressful life events (e.g. moving

TABLE 7.7 *Dataset 7.2*

events	status	stress	events	status	stress
4	0	23	68	1	15
10	0	25	77	0	32
7	1	19	15	0	24
10	1	9	74	0	32
10	0	18	69	0	23
29	0	33	45	0	28
12	1	17	28	0	19
6	0	20	95	0	26
14	1	21	21	1	19
62	1	8	5	1	20
70	0	24	72	1	26
70	0	27	66	1	18
19	1	16	94	0	34
43	0	24	16	1	23
10	0	12	34	0	27
14	1	15	61	1	30
61	1	24	4	0	20
22	0	14	77	1	19
23	1	19	52	0	19
35	1	13	92	0	31
56	1	21	50	1	19
93	0	35	33	1	26
35	1	17	13	0	20
51	1	17	53	0	36
26	0	25	58	1	19
32	1	26	35	0	17
5	1	20	98	0	27
54	0	22	73	0	27
72	1	9	87	1	21
41	1	17	24	1	7

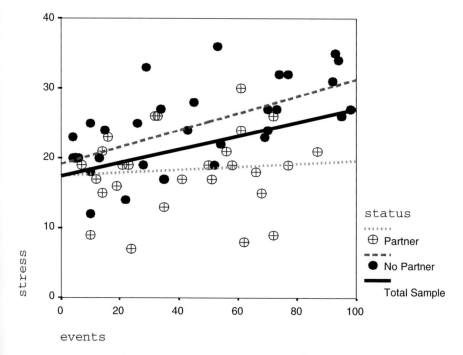

FIGURE 7.4 *Scatterplot of* events *versus* stress, *showing the lines of best fit for whole sample,* status = 0 *and* status = 1

house, changing job), status is a measure of whether a person co-habits with a partner (0 indicates that they do not, and 1 indicates that they do), and stress is the score on a self-report measure of experienced stress.

The psychologist is interested in testing the following three hypotheses:

1. Higher life events scores will lead to higher levels of stress.
2. People with a partner will feel less stress overall than people who do not have a partner.
3. Having a partner will reduce the impact of stressful life events – the increase in stress because of life events will be reduced in participants who have a partner.

One way to start exploring the data would be to represent the data on a scatterplot showing the separate slopes for those with a partner and those without, as in Figure 7.4. This chart shows that there does seem to be a relationship between events and stress as the line of best fit for the total sample slope upwards. Additionally, the line of best fit for the group with a status of zero is lower than the line for the group where the status is one.

Another way to start exploring the data would be to carry out a regression analysis for each group, and then to compare the slope and intercept parameters for the two groups. The summary statistics are shown in Table

7.8. It can be seen that for the group that has no partner (0) there is a strong association between events and stress. The predictor variable explains almost 37% of the variance in stress and this is significantly greater than zero. However, for the group that has a partner there is no association between events and stress. The result is not statistically significant (and adjusted R^2 is actually negative).

TABLE 7.8 *Summary of separate regression analyses for* status = 0 *and* status = 1

Status	R	R^2	Adj. R^2	F change	df 1	df 2	Sig. F
No partner	0.623	0.388	0.366	17.742	1	28	0.000
Partner	0.098	0.010	−0.026	0.270	1	28	0.608

Table 7.9 shows the parameter estimates for each of the groups. It can be seen that the slope for the 'no partner' group is strong and statistically significant, whereas the slope for the 'partner' group is not statistically significant.

TABLE 7.9 *Parameter estimates for separate analyses for* status = 0 *and* status = 1

Status	IV	Parameter estimate	Std error of slope	Standardised slope (beta)	t	Sig.
No partner	Constant	19.157	1.610		11.8	<0.001
	events	0.121	0.029	0.623	4.2	<0.001
Partner	Constant	17.482	1.926		9.0	<0.001
	events	0.02	0.041	0.098	0.5	0.608

This table shows that the slope is higher than for those that have no partner, but this difference in the slopes does not tell us whether the difference in the slopes is *significantly* different. It is possible that the slopes are both near to the 0.05 cut-off level for determining statistical significance: it could be that one is just above the level of significance, and the other is just below (although this scenario is unlikely in this case). It is also possible that neither slope is significant, or that both may be significant, and there still may be a significant difference between them. The same problem occurs for the constant parameters. The difference in the two constants tells us about the difference between the predicted levels when events is zero. If we examine the chart in Figure 7.4, we see that nobody has actually scored zero on the life events measure, and since the slopes are not parallel, the difference between them changes at different values of the predictor variable.

Hence we need a different approach to test adequately the third hypothesis (having a partner will reduce the impact of stressful life events − the increase in stress because of life events will be reduced in participants who have a partner). The approach which we need to use is very similar to the one we used when we analysed two categorical variables. The status variable is already currently treated as being dummy coded (see Chapter 3). We need to create an interaction term between status and events to capture the slope difference (much as we did previously with the two categorical predictors). However, first we need to carry out a procedure called 'centring' on the variable events. To do this we subtract the mean score of the variable events from each participant's score on that variable. Effectively we are transforming each person's score into a deviation from the mean. This transformation 'centres' the variable around the mean of 0. The next step is to create an interaction term by multiplying the centred events variable by the status variable. We have labelled the product term sxe, following our labelling convention from the previous section. Table 7.10 shows the first 10 cases from the dataset with the centred events data (called c_events), and the interaction term, sxe.

TABLE 7.10 First 10 cases from dataset 7.2b

events	status	stress	c_events	sxe
4.00	0	23	−44.03	0.00
10.00	0	25	−38.03	0.00
7.00	1	19	−41.03	7.00
10.00	1	9	−38.03	10.00
10.00	0	18	−38.03	0.00
29.00	0	33	−19.03	0.00
12.00	1	17	−36.03	12.00
6.00	0	20	−42.03	0.00
14.00	1	21	−34.03	14.00
62.00	1	8	13.97	62.00

Finally, a hierarchical regression is carried out, as described in Chapter 2. At the first step, the dummy variable (status) and the continuous variable (events) are entered into the equation. At the second step, the interaction term (sxe) is entered. The significance of the difference of R^2 represents the significance of the interaction effect. (A hierarchical regression is necessary because there are often problems with collinearity.)

Table 7.11 shows R, R^2 and Adj. R^2 for the model without the interaction term (Model 1) and with the interaction term (Model 2). The change in R^2 is equal to 0.042, with $F = 4.02$, and $df = 1, 56$, which is significant at $p = 0.05$. The interaction effect is therefore significant, and we can conclude that the slopes differ.

TABLE 7.11

Model	R	R^2	Adj. R^2
1 (No interaction term)	0.606	0.367	0.345
2 (Interaction term)	0.640	0.409	0.378

The parameter estimates are presented in Table 7.12. We can use these to construct a regression equation for each group.

TABLE 7.12 *Coefficients from regression equation*

	Slope (b)	Std error of slope	Standardised slope (beta)	t	Sig.
Constant	24.973	0.955		26.136	<0.001
events	0.121	0.031	0.519	3.942	<0.001
status	−5.964	1.361	−0.454	−4.382	<0.001
sxe	−0.099	0.050	−0.263	−2.004	0.050

The regression equation is as follows:

$$\text{stress} = b_1 \times \text{events} + b_2 \times \text{status} + b_3 \times \text{sxe} + c$$
$$\text{stress} = (0.121 \times \text{events}) + (-5.964 \times \text{status}) + (-0.99 \times \text{sxe}) + 24.973$$

We can expand this equation, by substituting the actual values that were used to calculate sxe:

$$\text{stress} = (0.121 \times \text{events}) + (-5.964 \times \text{status}) + (-0.99 \times \text{status} \times \text{events}) + 24.973$$

For the group without partners, in which status = 0:

$$\text{stress} = (0.121 \times \text{events}) + (-5.964 \times 0) + (-0.99 \times 0 \times \text{events}) + 24.973$$

When we multiply by the zeros:

$$\text{stress} = (0.12 \times \text{events}) + 24.973$$

Remember, though, that the score we used was a centred version of events, and therefore we need to correct for this centring. We created the new events variable by subtracting 48.03 from the previous variable. This means that the intercept in this equation will be too high by 48.03 × b_1. If we substitute the value of b_1, the expression becomes:

$$43.08 \times 0.121 = 5.170$$

The intercept is therefore given by:

$$24.973 - 5.170 = 19.803$$

Therefore, our final equation is:

$$\text{stress} = (\text{events} \times 0.12) + 19.80$$

Now if we do the same for the group with partners:

$$\text{stress} = (0.121 \times \text{events}) + (-5.964 \times 1) \\ + (-0.99 \times 1 \times \text{events}) + 24.973$$

when we multiply for status:

$$\text{stress} = (0.121 \times \text{events}) + (-5.964) + (-0.099 \times \text{events}) + 24.973$$

We can combine the two terms that are simply numbers (24.973 − 5.96 = 19.043) and the two terms that include events (events × 0.12 + events × (−0.099) = events × 0.021) to give:

$$\text{stress} = \text{events} \times 0.021 + 19.043$$

Again we need to correct for the fact that we centred the events variable by subtracting 43.08. To do this we add 43.08:

$$\text{stress} = \text{events} \times 0.021 + 19.043 \\ \text{stress} = (43.08 \times 0.021) + 19.043 \\ = (0.906) + 19.043$$

and:

$$0.906 + 19.043 = 19.949$$

Therefore the final equation for the group with partners is:

$$\text{stress} = \text{events} \times 0.021 + 19.949$$

Going all the way back to Table 7.12, the slope for status is −5.96. When we had a dummy variable with no interaction effects, in Chapter 3, the slope represented the difference between the means of the two groups. Now there is a problem because the lines are not parallel, so the difference between the two groups varies at different levels of events. The parameter could represent the difference in the two groups at the intercept (i.e. events = 0), but this would be a meaningless comparison, because nobody scored 0. Instead, the slope for status is associated with the difference between the groups at the mean level of events, which is a more meaningful point at which to make the comparison.

The procedure that we have described generalises readily to more complex situations, including the case where:

- the categorical variable may have three or more possible values;
- there may be other predictor variables that could be entered into the equation (for control purposes);

- there may be two or more categorical variables as well as a continuous predictor variable.

For further details on this type of situation, we suggest you read Jaccard, Turrisi and Wan (1990) or Aiken and West (1991).

7.2.3 Two continuous predictors

Finally, we describe the case where the relationship between a continuous predictor variable and a continuous outcome variable is moderated by a continuous variable. This means that the effect of the first continuous variable depends upon the level of the second continuous variable. For this example we return to the dataset in Chapters 1 and 2 where we examined the relationship between the number of books read, the number of lectures attended and the final grade students achieved in a module. The dataset is reproduced in Table 7.13.

In the analysis that we carried out in Chapter 2, we assumed that there was an additive relationship between the number of books read, rate of attendance at lectures and grade. That is to say, we assumed that the effect of reading each book was equivalent to reading any other extra book whether a student had attended one lecture or all of the lectures. Similarly we assumed that attending one more lecture would have the same effect, regardless of how many books had been read. These assumptions may have been incorrect: reading a book would be likely to have a different effect on the grade of a student who had not attended any lectures than on the grade of a student who had attended almost all of the lectures. The student who had not attended any lectures may be making a desperate attempt to improve their chances but reading an extra book may not be enough to make any difference.

To explore the data for the possibility of such interactions is rather more difficult than in the previous examples. One of the simplest ways to do it is to create a dichotomous variable from one of the continuous predictor variables (categorising attend as high or low around the mean or median) and then to employ the methods that we used in the previous section (in which we had a categorical moderator of a continuous variable). Note that by using a categorical moderator, we are discarding a great deal of information about the scores of the people; therefore we should not base any statistical tests on this analysis but we can use the technique to explore the possible form of any relationship.

If we split the attend variable at the mean thus creating two groups, the high attenders and the low attenders, we can then calculate the correlation between books and grade for each group:

- Low attenders: $r = 0.233$, $N = 19$, $p = 0.233$.
- High attenders: $r = 0.508$, $N = 21$, $p = 0.019$.

TABLE 7.13 *Dataset 2.1 (repeated)*

books	attend	grade
0	9	45
1	15	57
0	10	45
2	16	51
4	10	65
4	20	88
1	11	44
4	20	87
3	15	89
0	15	59
2	8	66
1	13	65
4	18	56
1	10	47
0	8	66
1	10	41
3	16	56
0	11	37
1	19	45
4	12	58
4	11	47
0	19	64
2	15	97
3	15	55
1	20	51
0	6	61
3	15	69
3	19	79
2	14	71
2	13	62
3	17	87
2	20	54
2	11	43
3	20	92
4	20	83
4	20	94
3	9	60
1	8	56
2	16	88
0	10	62

We find that for the high attenders, the correlation between books and grade is both positive and significant, whereas for the low attenders, the correlation between books and grade, although positive, is not significant.

Similarly, we can plot a scatterplot and draw a line of best fit for each of the groups (Figure 7.5) of high and low attenders. The result seems to be showing the same thing as we saw from the correlations: that is, the slope is higher for the high attenders than the low attenders. Many statistical packages can draw a three-dimensional scatterplot. Although 3-D scatterplots

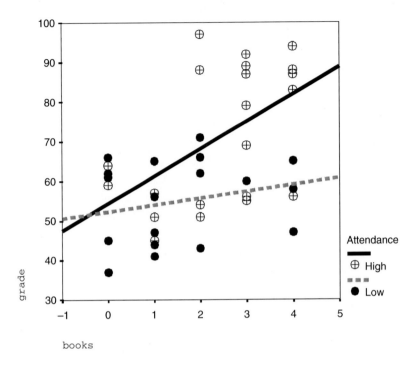

FIGURE 7.5 *Scatterplot showing line of best fit for high and low attenders*

may seem useful, they are difficult to interpret. Figure 7.6 shows such a scatterplot. We have added floor spikes to it, and also rotated the graph, to facilitate interpretation.

The graph is difficult to interpret, but, seems to show that, at the low end of attend, there is a slight increase in grade as the number of books increases. At the higher end of attend, grade starts higher (as we would expect) but there seems to be a larger increase in grade as books increases.

Clearly a visual inspection is not adequate to understand fully the relationships present in the data — we need to use regression for the data analysis. To analyse these data appropriately, we need to create a variable to represent the interaction term. This analysis is carried out in the usual fashion, by creating a product term between the two predictor variables. However, first we need to standardise the variables (also known as converting the variables to *z*-scores). Many statistical packages standardise variables automatically, but it is easy to do by hand. A standardised score (or a *z*-score) represents the number of standard deviations from the mean of that variable. To calculate a *z*-score, the mean of the variable is first subtracted from a score and then it is divided by the standard deviation of the variable.

The means and SDs are shown in Table 7.14. To create the *z*-score of attend, which we call zattend, we calculate:

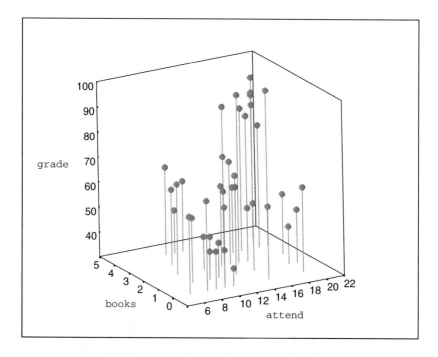

FIGURE 7.6 *Three-dimensional scatterplot of* books, attend *and* grade

TABLE 7.14 *Mean and SD of* books *and* attend

	Mean	SD
books	2.00	1.43
attend	14.10	4.28

$$\text{zattend} = \frac{\text{attend} - 14.10}{4.28}$$

Similarly we calculate the *z*-score of books using:

$$\text{zbooks} = \frac{\text{books} - 2.00}{1.43}$$

Finally we calculate an interaction term. Following our previous terminology, we refer to this as bxa, and it is calculated by multiplying the two standardised variables together:

$$\text{bxa} = \text{zbooks} \times \text{zattend}$$

We have reproduced the first 10 rows of the dataset in Table 7.15. Note that we have shown the values to three decimal places only, although most statistical packages will store many more decimal places.

TABLE 7.15

books	attend	grade	zbooks	zattend	bxa
0	9	45	−1.396	−1.192	1.665
1	15	57	−0.698	0.210	−0.147
0	10	45	−1.396	−0.958	1.338
2	16	51	0.000	0.444	0.000
4	10	65	1.396	−0.958	−1.338
4	20	88	1.396	1.379	1.926
1	11	44	−0.698	−0.725	0.506
4	20	87	1.396	1.379	1.926
3	15	89	0.698	0.210	0.147
0	15	59	−1.396	0.210	−0.294

As in the previous examples the data are now entered into a hierarchical regression. However, it is simpler to interpret the results if we enter the variables in their standardised form rather than their raw score form. Using raw score values does not affect the standardised values or the significance values, but using the raw scores will alter the standardised slopes. R, R^2 and adjusted R^2 for model 1 (no interaction term) and model 2 (interaction term added) are shown in Table 7.16.

TABLE 7.16

Model	R	R^2	Adj. R^2	F	df1	df2	Sig.
1	0.573	0.329	0.292	9.059	2	37	0.001
2	0.634	0.402	0.352	9.073	3	36	<0.001

The difference in R^2 is equal to 0.073, which has an associated F of 4.43, $df = (1, 36)$ and $p = 0.042$. We can therefore conclude that the interaction is significant. The slope coefficients are shown in Table 7.17. The unstandardised slope parameters for zbooks and zattend now represent the change in the dependent variable associated with a change of one standard deviation in the books and grade variables rather than a change of one of the original units.

We can interpret these slope coefficients in terms of the regression equation:

$$\text{grade} = b_1 \times \text{zbooks} + b_2 \times \text{zattend} + b_3 \times \text{bxa} + c$$

Substituting the values from Table 7.17 gives:

$$\text{grade} = (5.95 \times \text{zbooks}) + (5.70 \times \text{zattend}) + (4.50 \times \text{bxa}) + 61.60$$

As in the previous example, we can interpret this by expanding bxa to the terms that we used to calculate it (remember that bxa = zbooks × zattend):

TABLE 7.17

	Slope (*b*)	Std error of slope	Standardised slope (beta)	*t*	Sig.
Constant	61.602	2.319		26.570	0.000
zbooks	5.951	2.403	0.356	2.476	0.018
zattend	5.702	2.404	0.341	2.372	0.023
bxa	4.503	2.140	0.272	2.104	0.042

$$\text{grade} = (5.95 \times \text{zbooks}) + (5.70 \times \text{zgrade}) \\ + (4.50 \times \text{zbooks} \times \text{zattend}) + 61.60$$

We can now work out a predicted score for a person, given the number of books read and the number of lectures they attended using this equation, by substituting the values as before.

More interestingly, if we now substitute different values of attend, we can find the intercept and slope at those different values of attend. If we are interested in the increase in grade associated with reading one more book for a person who has attended five lectures, we first need to calculate the *z*-score associated with five lectures.

To do this we repeat the calculation we did to transform the value of attend into a *z*-score:

$$\text{zattend} = \frac{\text{attend} - 14.10}{4.28}$$

$$= \frac{6 - 14.10}{4.28}$$

$$= -1.89$$

To find the intercept and slope for zattend = −1.89, that is where attend is equal to six lectures, we substitute the value of −1.89 into the following regression equation (we have added some brackets to show the three parts of the equation more simply — they are not necessary):

$$\text{grade} = (5.95 \times \text{zbooks}) + (5.7 \times \text{zattend}) \\ + (4.5 \times \text{zbooks} \times \text{zattend}) + 61.6$$

When we substitute the value −1.89 for zattend we find:

$$\text{grade} = (5.95 \times \text{zbooks}) + [5.7 \times (-1.89)] \\ + [4.5 \times \text{zbooks} \times (-1.89)] + (61.6)$$

We can simplify this to:

$$\text{grade} = (5.95 \times \text{zbooks}) + (-10.773) + (-8.5 \times \text{zbooks}) + 61.6$$

We can now add together the two terms that contain only numbers and the two terms that contain zbooks, to give (we have now removed the brackets):

$$\text{grade} = (-2.55) \times \text{zbooks} + 50.8$$

We are using the variable zbooks, which is the standardised form of the variable books — hence a one-unit change in zbooks is associated with a change in books of one standard deviation, or 1.43 books. To find the change parameter associated with one book, we must divide the coefficient associated with zbooks by the standard deviation (1.43):

$$\text{Books coefficient} = -2.55/1.43 = -1.78$$

Therefore, if a student attends six lectures only:

$$\text{grade} = \text{books} \times (-1.78) + 50.8$$

Although it may seem curious that this value is equal to a negative number, it is sufficiently close to zero not to be significantly different from it. Therefore, we can conclude that, if a student has only attended six lectures, reading additional books will not help the student to achieve a higher grade.

We can calculate the slope and intercept value for every value of lectures to find the increase in grade associated with reading one more book. Table 7.18 shows the value of the grade change associated with reading one more book for all values of attend, from 6 to 20, and Table 7.19 shows the grade change associated with attending another lecture at different levels of books.

TABLE 7.18 *Grade increase associated with reading one more book, at different levels of lectures attended*

Lectures attended	Grade change
6	−1.79
7	−1.06
8	−0.32
9	0.41
10	1.15
11	1.88
12	2.62
13	3.35
14	4.09
15	4.82
16	5.56
17	6.29
18	7.03
19	7.76
20	8.50

TABLE 7.19 *Grade increase associated with attending one more lecture, at different levels of books read*

books	Grade change
0	−0.14
1	0.60
2	1.33
3	2.07
4	2.80

The moral of this story is (as all your tutors could have told you) that it is not sufficient only to read books, and it is not sufficient only to attend lectures. To do well, you must both attend lectures *and* read textbooks.

You may have noticed that this analysis is very similar to the analysis that we carried out in the section in Chapter 6 where we examined quadratic curve fitting. In fact, when there is high collinearity between two variables, the quadratic effect and the moderator effect may be very similar to one another (MacCallum and Mar, 1995).

Interactions with continuous variables can take on more complex forms, and can include non-linear terms and higher order interaction terms. These more complex forms are covered in some detail in Aiken and West (1991), and Jaccard, Turrisi and Wan (1990).

7.3 Mediator analysis

As well as acting as a moderator of the relationship between two variables, a variable may act as mediator of the relationship between a predictor and an outcome variable. A variable is said to mediate the relationship between a predictor and an outcome variable if the predictor variable first has an effect on the mediator variable, and this in turn influences the outcome variable.

The effect of sex on height is a mediated relationship. People who are genetically male tend to be taller than females, but this effect is not caused directly because of their genes, rather it is mediated by the influence of hormones. We can draw the relationship in a path diagram, as shown in Figure 7.7.

In the previous example, hormones mediated the relationship between sex and height to such an extent that genes had no influence on height other than the influence through hormones. In this case, the effect of hormones is

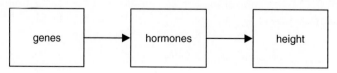

FIGURE 7.7 *Hormones as a mediator of the relationship between sex and height*

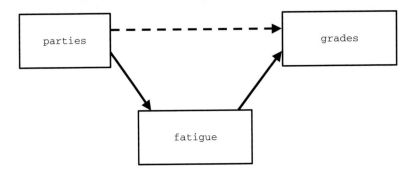

FIGURE 7.8 *The relationship between parties, fatigue and grades*

referred to as a *complete* mediator. Sometimes a variable may be a 'partial mediator' of the relationship between a predictor variable and an outcome — the predictor variable exerts some of its influence via a mediating variable, and it exerts some of its influence directly and not via a mediator.

Consider Figure 7.8, which shows the relationship between the number of parties attended by students, their level of fatigue and the grades they achieve. Students who go to many parties are likely to be tired at college the next day. This tiredness will affect their grades, as they will not be able to concentrate in lectures and seminars. The solid black arrow in the figure shows the relationship between parties, fatigue and grades. However, parties will have another influence on grades: students who are at parties are not writing essays, reading books or researching in the library. Thus the number of parties students attend will have an effect on grades which is not mediated by fatigue. Fatigue is a partial mediator of the relationship between parties and grades.

Baron and Kenny (1986) wrote a very important article that described the detection of mediator relationships. They describe four steps that must be taken to establish that a mediated relationship exists. Consider three variables, X, the predictor variable, Y, the outcome variable and M, the mediator variable. The steps are:

1. Show that X is a significant predictor of Y, using regression.
2. Show that X is a significant predictor of M using regression.
3. Show that M is a significant predictor of Y, when we control for X. To do this we carry out a multiple regression using X and M as predictors, and Y as the outcome.
4. If M is a complete mediator of the relationship between X and Y, the effect of X, when controlling for M, should be zero. If it is only a partial mediator the effect will be merely reduced, not eliminated.

7.3.1 Example of mediation

How much people enjoy reading books is likely to be a strong influence on the number of books they buy, and the number of books they buy in turn will

MODERATOR AND MEDIATOR ANALYSIS

TABLE 7.20 Dataset 7.3

enjoy	buy	read
4	16	6
15	19	13
1	0	1
11	19	13
13	25	12
19	24	11
6	22	7
10	21	8
15	13	12
3	7	4
11	28	15
20	31	14
7	4	7
11	26	14
10	11	9
6	12	5
7	14	7
18	16	12
8	20	10
2	13	6
7	12	9
12	23	13
13	22	9
15	19	13
4	12	9
3	10	5
9	7	7
7	22	8
10	7	8
2	0	2
15	16	7
1	17	6
3	11	9
6	5	9
13	29	15
15	29	11
16	20	9
14	16	7
1	3	2
8	8	10

be a strong influence on the number of books that they read. However, it is possible that books bought will not be a complete mediator of the relationship between how much people like books and how many books they read. People who enjoy reading books may use libraries or borrow books from friends.

Table 7.20 shows a dataset that contains three variables: enjoy represents a score on a scale designed to measure how much people enjoy reading books; buy is a measure of how many books people have bought in the previous 12 months; `read` is a measure of the number of books that people have read.

We will now work through the four steps described by Baron and Kenny (1986), using `enjoy` as the predictor variable (X in our previous terminology), `read` as the outcome variable (Y), and `buy` as the mediator variable (M).

Step 1: Show that X is a significant predictor of Y. To do this we must carry out a regression using `read` as the dependent variable and `enjoy` as the dependent variable. When we do this, we find that the slope coefficient is equal to 0.487 (standardised slope coefficient is equal to 0.732), which is significant at $p < 0.001$. The first condition of mediation is satisfied.

Step 2: Show that X is a significant predictor of M using regression. To do this, we need to carry out a regression analysis using `buy` as the dependent variable and `enjoy` as the dependent variable. When we do this analysis, we find that the slope coefficient is equal to 0.974 (standardised slope is equal to 0.643), $p < 0.001$, and thus the second criterion for mediation has been satisfied.

Step 3: Show that M is a significant predictor of Y, when we control for X. To carry out this step, we use `read` as the dependent variable, and both `enjoy` and `buy` as independent variables, and then examine the slope coefficient for `buy`. This is equal to 0.206 (standardised slope is equal to 0.469), $p < 0.001$. The third step for finding mediation has been satisfied.

Step 4: If M is a complete mediator of the relationship between X and Y, the effect of X, when controlling for M, should be zero. If it is only a partial mediator, the effect will be merely reduced, not eliminated. We use the same analysis as in the previous step, but now examine the slope coefficient for `enjoy`. This is now equal to 0.287 (standardised slope is equal to 0.431), $p = 0.001$. We cannot, therefore, conclude that `buy` is a complete mediator of the relationship between `enjoy` and `read`. To do this we would have to find this relationship to be zero, or at least not significant. However, the slope has been reduced from 0.487 to 0.287, and we can therefore conclude that partial mediation has occurred.

The amount of mediation is calculated by finding the difference in the slopes we found in steps 1 and 4, which is $0.487 - 0.287 = 0.200$.

7.4 Some concluding points on moderation and mediation

It is possible for moderation and mediation to combine in ways that are more complex. This can happen in two ways, referred to as moderated mediation, or mediated moderation. It is also possible for multiple independent variables to be mediated in their effects by one or more variables. Finally, mediator effects may also be non-linear.

Note

1 Actually, we are guilty of oversimplifying a rather complex issue here. The robustness of ANOVA to violations of assumptions is altered in unbalanced designs (see e.g. Maxwell and Delaney, 1990).

Further reading

Moderators are covered in two books devoted to the topic: Jaccard, Turrisi and Wan (1990) *Interaction Effects in Multiple Regression* and Aiken and West (1991) *Multiple regression: testing and interpreting interactions.*

Mediators are covered in Cohen and Cohen (1983) *Applied multiple regression/correlation analysis for the behavioral sciences* (2nd ed.), Chapter 9.

8 Introducing some advanced techniques: multilevel modelling and structural equation modelling

In this final chapter we will briefly cover two techniques which have their roots in regression analysis, but which involve more advanced forms of analysis: multilevel modelling and structural equation modelling. These techniques can be used to ask questions of your data that would not be possible using the standard regression techniques we have described so far. The information presented here should be treated as a very brief overview and for those who wish to pursue the subject, there is further reading that we recommend. These techniques often cannot be carried out with many of the more common statistical analysis packages; at the end of the chapter, we will have a brief look at some of the software that can do these types of analysis.

8.1 Multilevel modelling (MLM)

It is often said that necessity is the mother of invention, and MLM is an example of an analysis technique that was developed for a specific need. In many types of study we collect data from people who are grouped into clusters. In this case measurements from these people will not be independent, and hence we will violate the assumption of independence, discussed in Chapter 4. There, we said that every case should be equally related (or unrelated) to every other case. MLM was developed, and was originally used in school-effectiveness studies. Pupils in schools are clustered into classes, classes are clustered in schools, schools are clustered in areas, areas in districts and districts in countries. Two pupils who are in the same classroom will be more similar to each other than two pupils who are in different classrooms; similarly two classrooms in the same school will be more similar to one another than two classrooms in different schools, and so on. Data that have this type of structure are referred to as multilevel data, or hierarchical data. MLM is sometimes called hierarchical linear modelling (HLM). Some books also refer to 'random slope modelling', or 'mixed modelling'.

Studies in educational effectiveness examine the effects on achievement of factors such as class size, teaching style, or sex segregation. In such a study, we may wish to control for initial ability. In the simplest form, a school-effectiveness study would have one outcome variable (usually examination performance), and two independent variables: a measure of initial ability, taken when the pupils start at the school, and a measure of the variable in which we are interested, for example teaching styles. Deciding upon the most appropriate level of analysis is a problem when using the regression analysis that we have described in the rest of this book to analyse data that have a multilevel structure.

If we were to analyse the data using a usual regression approach, we would analyse individuals, who would each make one case in our analysis. This is usually called the first level, or Level 1. If we were to carry out this analysis, at Level 1, the assumption of independence required by regression models is violated and this violation of the assumption of independence may lead to inflated relationships and a higher type I error rate, as was discussed in Chapter 4. If higher level units are analysed (e.g. classes (Level 2) or schools (Level 3)), then information regarding the actual score of each pupil is discarded. Discarding this information may therefore reduce the measured relationships between variables, and inflate the type II error rate (Goldstein, 1995). A practical example of a type I error was demonstrated by Aitkin et al. (1981). They reanalysed an important and influential study into teaching styles carried out by Bennett (1976). In the original study, which had used a regression-based approach, a significant difference was found between teaching styles. However, when Aitkin et al. reanalysed the data accounting properly for the multilevel structure of the data, the significant differences disappeared.

It is also important to analyse the data at an appropriate level, because the relationship at Level 1 and the relationship at Level 2 may not be the same, and it is very tempting to generalise from one level to another, where this may be inappropriate.

Figure 8.1 shows a (fictional) graphical example of how a failure to account for the multilevel nature of a dataset can lead to a situation where a negative relationship at Level 1 co-exists with a positive relationship at Level 2. When this occurs, generalising from one level to another level can lead to a faulty interpretation, and so we must be especially careful to know which level we are talking about. Students worked on one of three statistical problems, labelled as w, x and y; the time they spent working on the problem and the grade they achieved were measured. These data are hierarchical, the Level 1 units being the individual students, and the Level 2 units being the problems they worked on. Each of the letters (ignoring for now what the letter actually is) in Figure 8.1 represents a student (Level 1 unit). Each group of letters (w, x, y) represents a Level 2 unit (a problem). The apparent relationship between the two variables, hours and grade, when we analyse these data at Level 1 is represented by the thick black line, which shows a positive relationship – more hours worked leads to higher grades. The cluster of letters surrounding each line represents the hierarchical nature of the data. But you can see that a negative relationship between the number of hours worked and the grade achieved exists *within* each problem. Thus if we ignore the hierarchical nature of the data, a negative relationship between hours and grade has been disguised, and may be interpreted as a positive relationship. The relationship at Level 1 and the relationship at Level 2 are different, and if we are to talk

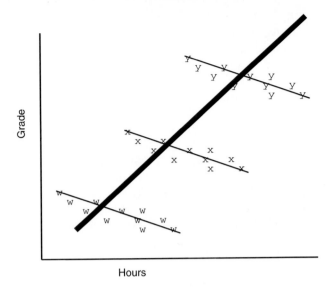

FIGURE 8.1 *Reversal of direction of correlation*

about the relationship between hours worked on a problem and the grade students achieve, we must be careful not to generalise from one level to another. Problems that took longer gave rise to higher grades, because they were harder problems, and took more effort to answer. Students that took longer on an individual problem got a lower grade, because they were less practised at the problem, and made more errors.

MLM therefore has two advantages: the first should be clear, that is we are specifying the level at which the relationship exists, and not generalising to an inappropriate level. The second advantage relates to power in the higher level units. In this example, 11 students attempted each problem. If we analyse at Level 1 (inappropriately) using regression, we have $N = 33$. If we wanted to analyse the data more appropriately at Level 2 using regression, we would need to aggregate the data within each problem. So for Problem 1 we would find the mean time taken and the mean grade, similarly for Problem 2, and similarly for Problem 3. We would now look for a relationship between these two variables, but we have thrown away a massive amount of information, because now, rather than having a sample size of 33, we have a woefully inadequate sample size of three. By using MLM, we retain all of the information that was present in the data while analysis is carried out at each level.

8.1.1 Algebraic formulation

We have been telling you most of the way through this book that the equation for a regression calculation is:

$$y = b_1 x_1 + c + e$$

However, you will remember from Chapter 1 that this was a simplification. A more general example would be:

$$y_i = b_1 x_{1i} + c + e_i$$

where y_i is the score of the ith person on the dependent variable (so where $i = 1$, y_i is the score of the first person), c is the intercept, b_1 is the estimate of the slope of the regression line, x_i is the score of the ith person on the independent variable, and e_i is the deviation from the predicted score for that person (i.e. the residual).

In simple MLM, we have i Level 1 units (individuals) nested within j Level 2 units (classes). Then, we introduce a new component into the equation, u_j. This represents the departure from the constant for each group. Now the equation becomes:

$$y_{ij} = b_1 x_{1i} + c + u_j + e_i$$

This equation tells us that the score for the ith person in the jth group is equal to:

- $b_1 x_{1i}$: the slope coefficient, multiplied by the dependent variable; plus
- c: a constant; plus
- u_j: a departure from the constant for each group; plus
- e: a residual for each person.

We can represent this equation graphically, as shown in Figure 8.2, where we use an example that has four Level 2 units (e.g. four different classes). The slope coefficient is equal to the increase in the dependent variable when the independent variable increases by one unit. This slope coefficient is shown in the figure as b, which is equal to 1.8. Because all of the slopes are equal we can use any of them to calculate b.

The constant is the average point at which the slopes intercept with the y-axis. The four different slopes all intercept the y-axis at a different point, but the average of all the lines is shown by the thick black line which crosses the y-axis at 3; therefore $c = 3$.

The values for u_j are the amount by which the intercepts for each slope differ from the intercept for the average slope. We have four groups, so we have four values for u_j. To find u_1 we see how far the intercept for Group 1 is above the average intercept (c). We can see in the graph that the intercept for Group 1 is equal to 5, since $5 - 3 = 2$, $u_1 = 2$. Similarly $u_2 = 1$, $u_3 = -1$ and $u_4 = -2$.

We can build up the model in a number of different ways. In this model we have forced all groups to have equal slopes, but we can allow slopes to vary across groups. We can also include more independent variables, and these can act at different levels: teaching style would act at the class level, whereas intelligence would act at the individual level. We can cluster the groups into

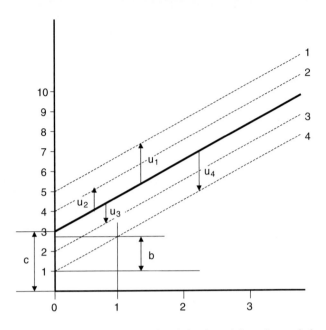

FIGURE 8.2 *Graphical representation of multilevel model, with equal slopes and varying intercepts*

higher level units, for example schools may be clustered into educational authorities, or districts; slopes can vary within these higher level units and independent variables can also act within them (see Woodhouse (1996) for details).

At higher level units we can examine the relationship between parameters that were estimated in lower level units. Figure 8.3 shows a two-level model where children's achievement was measured when they started at a particular school (Test 1) and then assessed again at a later date (Test 2). The slopes represent each school, and although the slopes differ they are all positive. More importantly, the schools with lower intercepts seem to have steeper slopes; this means that there is a negative correlation between the slope and the intercept (imagine working that out with standard regression analysis!). This shows that in the schools where the children start off as low achievers, the children seem to improve a great deal, whereas in schools where the children start off as high achievers, the children do not seem to improve as much.

8.1.2 Hierarchies everywhere

One of the problems that people find when they discover that multilevel (or hierarchical) data exist, is that suddenly everything becomes an MLM problem. As Kreft and De Leeuw (1998) put it, 'Once you know that hierarchies exist, you see them everywhere' (p. 1). A few examples are described in this section.

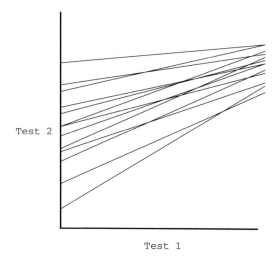

FIGURE 8.3 *Negative correlation between slopes and intercepts*

Many studies in social psychology involve the analysis of group behaviour. But each of the individuals in each of these groups will likely be more similar to their fellow group members than they are to people in different groups, and data on groups should therefore be treated as hierarchical data. Studies in clinical psychology and medicine often involve the analysis of patients taken from multiple centres – several hospitals, or clinics, for example. Each of these should be treated as a Level 2 unit. Research in occupational psychology can involve the analysis of relationships between variables measured on people who are clustered within departments, departments which are clustered within organisations, and organisations which may be clustered within types of industries. Each of these should be treated as a hierarchy.

8.1.3 Even more hierarchies

Many other types of data structure which could be analysed using 'traditional' techniques can be treated by MLM: the analysis may be equivalent, but the multilevel approach can be much more flexible. For example, in a longitudinal experiment, which would usually be analysed as a repeated-measures MANOVA of some kind, the data on the individuals can be treated as Level 2 data, and the individual measurements from each individual can be treated as Level 1. In the same way that pupils can be clustered within classes, and classes within schools, measures can be clustered within individuals. This approach provides a great deal more flexibility than in the usual approach, particularly when measurement of time intervals is being considered. In a MANOVA-based analysis, if a person drops out of the study for any reason that individual usually must be removed from the study. In a study based on the multilevel approach, an individual dropping out of the study is simply treated as would be a class with fewer pupils than another class. In addition the length of time the individual did participate in the study can be treated as an independent variable (at Level 2), and can be used as a predictor, in much the same way that a class size can be used as a predictor. A common problem in longitudinal research is determining whether those participants who drop out of a study are a random selection from all participants, or if a variable that we are interested in is causing them to drop out. More importantly, we want to know if the relationship between our variables differs between the participants who drop out and those who remain in the study. Whether participants who drop out are different is relatively easy to determine using a multilevel approach: the number of times an individual participated in the research programme can be treated as an independent variable. By looking for the main effects of this independent variable, we can determine whether the dropouts differ on the outcome variable. If we look for interactions between this independent variable and other variables, we can determine whether relationships between the variables differ for dropouts and non-dropouts.

8.2 Structural equation modelling

Structural equation modelling (SEM) is a general data analytic technique with a number of advantages over traditional approaches to statistical analysis and it is increasing in popularity in psychological research. SEM encompasses everything that 'traditional' regression techniques can accomplish, with lots more besides. The effectiveness of SEM is being recognised increasingly by researchers from many disciplines, and the use of SEM techniques has been increasing. Tremblay and Gardner (1996) documented the increasing use of SEM in psychological research and reported that the percentage of articles involving SEM had doubled between 1987 and 1994, and that there had been a similar increase in the number of journals which publish papers involving SEM.

We will first discuss why SEM is used and then introduce some basic concepts that are important to SEM.

8.2.1 Why use SEM?

SEM produces results equivalent to any that regression techniques can produce. In fact, regression analysis can be thought of as one particular type of structural equation model. (You may also remember from Chapter 3 that many other analyses can be thought of as types of regression, and this means that they can also be thought of as types of SEM.) So why, the alert reader asks, have we struggled through seven chapters of a book on regression when we should have learned about SEM from the start? And why have the authors wasted a reasonable proportion of their lives writing about regression when they could have saved the time by telling their readers to go and buy one of the excellent introductions to SEM? We offer two justifications:

1. SEM is difficult. The learning curve involved in SEM is reasonably steep, and most of what we have written in this book can be applied to SEM. To learn about SEM, it is first necessary to have a thorough understanding of regression analysis. It involves many issues which are controversial (and will continue to be controversial for the foreseeable future).
2. To attempt to learn about SEM without this very thorough understanding of regression analysis is not a task that we would care to inflict on anyone.

So if SEM is so difficult and has these unresolved issues, why are we writing about them at all? We have impressed upon the reader throughout this text that regression analysis is a very flexible approach which can be used in a wide variety of situations, and we now say that everything that can be done with regression analysis can be done with SEM. However, there are an even wider range of analyses that SEM analysis can do and regression analysis cannot.

SEM analysis has three main advantages over other forms of analysis and we will examine them in the following sections.

8.2.1.1 Tests of model fit

A regression analysis provides us with a set of parameter estimates and a set of standard errors for those parameter estimates. Because of the nature of the equations for the regression analysis, it is always possible to find satisfactory solutions to the equations. The model that we get from a regression analysis can never be 'wrong', in the sense of not fitting the data. (The model may be irritating, unsatisfactory or surprising, it may mean that we need to rethink our ideas about the structure of the processes underlying our measures, but it cannot be wrong.)

When we analyse data using an SEM approach, we formulate a hypothesis about the underlying model and we test that hypothesis. If the model is appropriate, we can interpret the parameter estimates. However, it is possible – even usual – for us to be wrong about the model. If the model is wrong, or (in SEM terminology) if the model does not fit the data, the parameter estimates will not be meaningful, and cannot be interpreted.

Determining whether a model is adequate in terms of fit is a complex problem, often hotly debated, so here we offer only a simple account of model fit.

8.2.1.2 Flexibility

We said in the previous section that SEM is a highly general analysis technique, and being so general makes it suitable for testing the types of complex hypotheses that are encountered in psychological theory. Often the theoretical processes that interest psychologists cannot be adequately described in terms of the narrow statistical models that we are restricted to in regression models. SEM allows us to test complex hypotheses that are more complex than those that were possible in regression.

In regression analysis we test parameters for significance by comparing them with zero. In SEM analysis we can test a parameter estimate against any value (including another parameter estimate), by asking 'Is the effect of X on Z significantly larger than the effect of Y on Z?' We can also compare parameter estimates between groups by, for instance, checking the answer to a question such as 'Is the total R^2 for Group 1 significantly different from the R^2 for Group 2?'

Things that SEM can deal with include: direct and indirect effects, interactions, parameters that are constrained to be equal (or to be a function of other parameters), reciprocal relations, and mean differences between groups within one specified model. Hoyle (1995) notes that ANOVA and exploratory factor analysis are generally not guided by psychological theory, whereas SEM is a theoretically driven procedure. As psychological theory becomes more advanced, more advanced models are needed to test hypotheses drawn from contemporary theory.

When using SEM we can be less concerned about model and distributional assumptions than we are when we are working with less general models. For example, in regression the assumption has to be made that the variables are free of measurement error. This assumption can be relaxed a little when these models are specified in SEM. Not having to make these assumptions provides conceptual and statistical advantages. The conceptual advantage is that the researcher can represent the psychological processes in a way that has better theoretical justification – that is, without biasing the estimated parameters by violating the distributional assumptions. Standard statistical approaches generally rely on continuous multivariate–normal data. Recent developments in SEM have proposed methods to deal effectively with dichotomous or ordinal data (e.g. Muthén, 1984) and violations of multivariate–normality assumptions (e.g. Browne, 1984; Satorra & Bentler, 1994). However, it should be noted that debate continues about how far these techniques can, and should, be used.

8.2.1.3 Error in measurement

Measurement error was discussed in Chapter 5, and probably the most powerful aspect of SEM is the ability to correct for measurement error. McNemar (1946) noted that 'all measurement is befuddled with error' (p. 294). In the physical sciences, measurement error is usually very small, and can often safely be ignored. ANOVA and related techniques were originally developed for use in agricultural research, where the dependent variables of interest were things like crop yield, or animal weight, and the independent variables included manure. Measurement error matters particularly to social scientists who are not generally interested in manifest, tangible variables like crop yield but in variables which are unobservable or latent – such as anxiety, stress or intelligence. The relationships between the observed variables, for instance the score obtained on a scale measuring anxiety, and the unobservable variable, the actual latent variable of anxiety will rarely, if ever, be perfect. In contrast, the correlation between the true weight of a number of pigs, and the measured weight of a number of pigs, will have a value very close to 1.00. So the observed measures of anxiety contain variance that is attributable to measurement error as well as to actual anxiety. The variance of observed measures of the weights of pigs would contain very little variance that can be attributed to measurement error.[1] In the social sciences we cannot measure variables with the same degree of accuracy as can the physical sciences: we are forced to act as if we are agricultural researchers whose weighing machine is inaccurate. There is an extensive literature describing the deleterious effects of measurement error (see Bollen (1989) for comprehensive coverage of the consequences of measurement error).

8.2.2 Identification

In statistical analysis, when we refer to the state of *identification* of a model, we are saying whether the equations in it can be solved with a unique solution. The process of calculating regression is the process of solving equations for the purpose of finding the values of unknown variables (the constant and the slope coefficients). Because these equations always lead to a unique solution, we describe these equations as *identified*.

Consider these much simpler examples:

$$x = 4$$

This equation is identified because there is only one possible solution, and we can find it. If, however, we have two unknown values we need more than one equation. The two equations:

$$x + y = 7$$
$$x - y = 3$$

can be solved with a unique solution, which is that $x = 5$ and $y = 2$, and therefore this system of equations is identified. However, consider the following equation:

$$x + y = 7$$

This equation does not have a unique solution because the equation cannot be solved, and therefore we say that it is *not identified*. Instead of having a unique solution, it has an infinity of possible solutions. The following solutions all work equally well and there is no way to decide between them:

- $x = 0, y = 7$
- $x = 7, y = 0$
- $x = 4, y = 3$
- $x = -2, y = 9$
- $x = 1\,000\,000, y = -998\,993$

In fact we could keep going for ever. To repeat, there are an infinite number of possible solutions to this problem, and hence it is not identified.

When there is one unknown value, we need one equation to find a unique solution. When there are two unknown values we need two equations to solve the equation. Such systems of equations are referred to as being *just identified*, because there are just enough equations to get one exact solution.

Sometimes we have more equations than we need for the number of unknown values that we want to find. Such systems of equations are referred to as being *over-identified*. The following set of equations is over-identified:

$$x + y = 6$$
$$x - y = 4$$
$$2x - 3y = 7$$

This set of equations can be solved using the values of $x = 5$, $y = 1$.

But now consider the following set of equations. This new set of equations cannot be solved at all – there are no values for x and y for which each of the equations holds true. They therefore cannot be a model of anything: they are inconsistent, and therefore they are wrong.

$$x + y = 6$$
$$x - y = 4$$
$$2x - 3y = 6$$

With three equations and two unknowns, it is possible for the equations to be inconsistent, to be wrong.

We can think of an equation as a piece of information that we know, a statement. Each equation (or mathematical statement) puts a piece of information into the system. A parameter estimate of x or y is something that we want to know, a piece of information that we take out of the system. If we put as many pieces of information into the system as we expect to take out, the system is, as we said, *just identified*. However, to take out more information than we put in is impossible: we cannot do it, and the system is *not identified*. (To get more out than we put in would be the statistical equivalent of getting something for nothing.) If we want to take out fewer things than we put in, then we are in profit, and the system is *over-identified*. And just as with your bank account, it is better to take out less than you put in.

We can think of regression analysis as a system of equations. When we have a regression analysis in which we have two independent variables (x_1 and x_2) and one dependent variable (y), we know six things which we put into the system:

1. Correlation of x_1 and x_2
2. Correlation of x_1 and y
3. Correlation of x_2 and y
4. Variance of x_1
5. Variance of x_2
6. Variance of y

When we estimate the regression equation, we estimate the following:

1. Slope for the effect of x_1 on y.
2. Slope for the effect of x_2 on y.
3. Correlation between x_1 and x_2 (although we do not report this, it is used to adjust the slope parameters).
4. Variance of x_1 (again, not reported and does not change).
5. Variance of x_2 again (not reported and does not change).
6. Variance accounted for in y (i.e. R^2).

So we put six pieces of information going into the analysis, and we take six pieces of information out at the other end. The analysis is therefore just identified. Because it is just identified, it can never be wrong.

An important aspect of an SEM system is that it is over-identified, so we take out less than we put in – we are in profit. And, just as in the real world, there are advantages to being in profit. SEM systems are (usually) over-identified, which means that models that are proposed in SEM can be wrong. And if we can specify our theory in terms of a model that can be wrong, our model, and hence our theory, can be tested.[2]

8.2.3 Latent variables

In psychology and most of the social sciences, we can never take a direct measure of the variable that we are interested in – we can only measure its effects. Sometimes psychologists make this distinction, and make it clear, but often they do not. We want to emphasise that the measured variable is not the same as the psychological variable that we are actually interested in. Some examples are as follows:

- Intelligence. What we really measure is the marks that people put on a piece of paper. We hope that these marks are caused by intelligence, but there are other factors that may also affect where people put the marks – guesswork, for example. We cannot measure intelligence *directly*, but we can measure its effects – and an intelligence test is a proxy for a measure of intelligence.
- Reaction time. Psychologists are not interested in reaction time *per se*. They are interested in the psychological process underlying that reaction time, and the reaction time is a proxy for measuring the time the underlying process took. Other things than the basic psychological process may

alter that reaction time – for example, if the participant is distracted for a moment.
- Eye movements. Psychologists recording eye movements are not actually interested in the eye movements, they are interested in the psychological process that is guiding the eye movements. They cannot measure those processes, so they measure the eye movements as a proxy.
- Psychologists may try to measure autonomic arousal to see how stressed participants feel at any time. They cannot measure autonomic arousal, so they use other measures as a proxy, for example heart rate, blood pressure, breathing rate, or galvanic skin response.

The variable that we are really interested in, that intangible abstraction that cannot be directly measured, is referred to as a *latent variable*. The thing we actually measure is called an *observed variable*.

Psychologists who research personality and individual differences tend to be familiar and comfortable with the idea that their measures are substitutes for the thing they really want to measure. In fact the idea was introduced by Spearman (of the correlation) in 1904, and it was popularised by psychologists like Thurstone, Cattell, Burt and Eysenck in the field of attitude and personality measurements. Psychologists from other fields do not always think of their variables as being proxies for the variable of interest.

Every psychological measurement has a *real* value that it is trying to measure. This is what we are trying to get at when we take reaction times, measure depression, etc. The real value, the one we can never quite get at, is referred to as the true score, and abbreviated to T. The measured score (x) is influenced by T, but also has some error (e) associated with it. We can therefore say that:

$$x = T + e$$

The score you get on an intelligence test has something to do with your true intelligence, but is also affected by factors like guesswork, luck, mood or level of fatigue. These factors are collectively called error (e).

Remembering what was discussed in the preceding section, you may notice that this equation is not identified. We have a single observed measure (x) which gives us one piece of information, but want to know T (the true score) and e (the error). Having only a single observed measure does not give us enough information to find unique estimates of values for T and e. However, if we have at least four measured variables, each of which is influenced by the latent variable, we have the following set of equations:

$$x_1 = T + e_1$$
$$x_2 = T + e_2$$
$$x_3 = T + e_3$$
$$x_4 = T + e_4$$

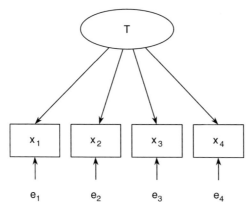

FIGURE 8.4 *A simple path diagram*

We can also represent this as a path diagram in which the measured variables are shown in boxes and the latent variable (*T*) is shown in an ellipse (Figure 8.4).

This path diagram specifies a specific model that postulates a latent variable that influences the observed measurements. The influences or effects (these are essentially slopes) are represented by the arrows. For example, if *T* is 'depression' the observed measurements may be loss of appetite (x_1), sleeplessness (x_2), low self-worth (x_3), and sadness (x_4). The model specifies a particular causal direction from the latent variable to the observed measures. Therefore it is hypothesised that an increase in depression leads to an increase in loss of appetite, an increase in sleeplessness, an increase in loss of self-worth, and an increase in sadness. As none of these indicators are perfect representations of depression, the relationships between the latent variable and the observed measures will not be perfect – the correlations will not be equal to 1.00. The unexplained variance in the observed measures is measurement error (*e*), also known as unique variance (variance that is unique to the latent variable).

In this example we have four variables, and these four variables combine to make 10 variances and covariances, as shown in Table 8.1.

TABLE 8.1

x_1	1			
x_2	2	3		
x_3	4	5	6	
x_4	7	8	9	10
	x_1	x_2	x_3	x_4

We do not need to draw this kind of table every time we want to know how many correlations there are between a set of variables, because we can use the formula:

$$\text{Number of elements} = \frac{k(k+1)}{2}$$

where k is the number of variables.

We now have 10 pieces of information (variances and covariances) and four things we want to know (the four path coefficients).[3] The model is now over-identified.

If the model were just identified we could merely work out the path coefficients for each measured variable. However, because it is over-identified, we have enough information to *test* the model. And if we can test the model, we can show it to be wrong.

This may seem like a curious anomaly. Why would anyone want their theory to be proved wrong? This is because, if a theory cannot be tested, it cannot be proved wrong, and if a theory cannot be proved wrong, it cannot be treated as a scientific fact (Popper, 1968). Many scientifically oriented psychologists consider the work of Freud to be unscientific, because there are no circumstances that would prove it wrong.

The model shown in Figure 8.4, in effect, says, 'I think that a psychological variable exists. And I think that the variable influences the score on four variables that we can measure directly.' If this is the case, the model will fit, but if it is not, the model will not fit. The existence of the latent variable can therefore be empirically tested by measuring the observable variables. We do not have simply to add up the variables or take the mean (as we would if it were a scale of some type), but we can assess the existence of the latent variable before we do so.

8.2.4 Estimation in SEM

In regression analysis we use the method of least squares to estimate the parameters in the model. In SEM, as in logistic regression (see Chapter 5), the least squares method will not work as it is not possible to determine a unique solution. Instead, as in logistic regression, we use *maximum likelihood* (ML) estimation.[4] ML estimation is an iterative technique that attempts to find the best solution by searching through solutions and testing them until it finds the best one. (This solution could be used in regression analysis – we hinted at it in Chapter 1 – as we could just try different values as parameter estimates until we find the values that minimise the sums of squares. This is not necessary in regression analysis, as we can use equations to find the best values.) As soon as we put some trial values for the unknowns into the equations we can work out the correlation matrix that would be implied by those values. This technique is usually done through matrix algebra (see Bollen, 1989), but in this simple model we can do it in a simpler way. The example in Table 8.2 shows a correlation matrix where four measures of autonomic arousal have been taken: GSR (Galvanic Skin Response) (x_1), heart rate (x_2), breathing rate (x_3) and blood pressure (x_4).

TABLE 8.2 *Correlation matrix of four measures of autonomic arousal*

x_1	1.00			
x_2	0.47	1.00		
x_4	0.49	0.36	1.00	
x_4	0.89	0.48	0.54	1.00
	x_1	x_2	x_3	x_4

We propose an initial solution to the equations. We suppose that the values we might find are:

- Path $x_1 = 0.4$
- Path $x_2 = 0.5$
- Path $x_3 = 0.6$
- Path $x_4 = 0.7$

We want to find the correlation matrix that would be implied by this model. We check to see if these numbers generate the 'answers' that we want. In this simple case we can find the implied correlation between two variables by multiplying the path coefficients for the two variables together. The implied correlation between x_1 and x_2 is given by 0.4×0.5, between x_1 and x_3 given by 0.4×0.6. We could carry on doing this and fill the values into an *implied* correlation matrix (Table 8.3).

TABLE 8.3

x_1	1.00			
x_2	0.20	1.00		
x_3	0.24	0.30	1.00	
x_4	0.28	0.35	0.42	1.00
	x_1	x_2	x_3	x_4

If we compare this matrix with the original sample correlation matrix, we can see that they are not very similar, and therefore we need to try again, with different values. The procedure attempts by a series of iterations to find the best solution for the four path coefficients by looking for the minimum value for the fit function (called F_{ML}). This is a similar process to the way in which the OLS estimate finds the values of the slope coefficients that minimise the sum of the squared residuals. Doing these calculations by hand would take an exorbitant amount of time, so we have used a computer program (LISREL 8.30; Jöreskog and Sörbom, 1999). The program found the following paths gave the best values for F:

- Path $x_1 = 0.91$
- Path $x_2 = 0.50$
- Path $x_3 = 0.56$
- Path $x_4 = 0.97$

The correlation matrix implied by this solution is shown in Table 8.4, along with the sample correlation matrix for reference.

TABLE 8.4 *Implied and sample correlation matrices*

	Implied correlation matrix				Sample correlation matrix			
x_1	1.00				1.00			
x_2	0.46	1.00			0.47	1.00		
x_3	0.53	0.28	1.00		0.49	0.36	1.00	
x_4	0.87	0.45	0.52	1.00	0.89	0.48	0.54	1.00
	x_1	x_2	x_3	x_4	x_1	x_2	x_3	x_4

We can see now that the correlation matrix implied by the values that were found using ML is very close to the values of the sample correlation matrix.

8.2.5 Model testing

We said that one of the advantages of using SEM is that the model can be tested, and is therefore falsifiable (able to be proved wrong). The value for F_{ML} can be used as a test of the null hypothesis that the sample and implied correlation matrices are the same. The value of F_{ML}, when multiplied by $N - 1$, is distributed approximately as χ^2. In SEM, the *degrees of freedom* are defined as the number of items in the covariance matrix minus the number of parameters in the model. The model has eight parameters, and we have 10 elements in the correlation matrix, so $df = 2$.

$$F_{ML} = 0.0279$$
$$\chi^2 = F \times (N - 1)$$
$$= 0.0279 \times (126 - 1)$$
$$= 3.48$$

If the value of χ^2 is very small, and produces a non-significant result, the implication is that the model is a good explanation, or *fits* the data. (It is one of the curious things about SEM that while in other forms of statistical analysis we usually hope for a probability value *less* than the chosen level of alpha, in SEM we hope for a probability value *greater* than the chosen level of alpha.)

The value of χ^2, with $df = 2$, has an associated probability value of 0.18. As this is greater than 0.05 (alpha) we can conclude that the model is an adequate description of the data.

We can add further restrictions to the model: for example, we can constrain path coefficients to be equal to one another. If we tell the ML procedure that we would like all four of the path coefficients to be equal to one another, we are hypothesising not only that the latent variable exists, but also that it has an equal effect on all of the measured variables. A model with these characteristics is called a parallel model. We are now effectively estimating only one path coefficient (since we have fixed them all to the same value). If we do restrain the path coefficients to be equal to one another, we find that the estimate of all four path coefficients is equal to 0.86. The value of χ^2 is now equal to 44.89, with $df = 5$, which has a probability of less than 0.001. We can therefore conclude that this model, in which all of the slopes are fixed to be equal, is not an adequate description of the real situation. The four measured variables are therefore not equally good measures of autonomic arousal.

8.2.6 Structural models

The small model we have considered may be interesting in its own right, but it is likely to be even more interesting as part of a larger model. We may make the hypothesis that when we ask people to do a frustrating task some types of them become extremely frustrated, and with the sort of person who becomes extremely frustrated, their degree of frustration increases their degree of autonomic arousal (i.e. they may get into a rage). Friedman and Rosenman (1974) have referred to the type of behaviour where frustration causes an unusual level of autonomic arousal as *Type A behaviour*. We can assess Type A behaviour using five items that assess different aspects of it by asking the subject these five questions:

x_1: Do you play every game to win (even against children)?
x_2: Are you always on time?
x_3: Do you become impatient while watching others do something you know you can do better?
x_4: Do you have difficulty sitting and doing nothing?
x_5: If you want something done well, do you have to do it yourself?

We want to test the hypothesis that these items are measuring the same latent variable. We can use the same method that we used when we tested this hypothesis for the autonomic arousal variable. Both models are referred to as measurement models – they look at the measurement of the latent variable. What is much more interesting is to combine these models, and then examine the relationships between the latent variables. This part of the model, which links two latent variables, is referred to as a structural model, and is shown in Figure 8.5. L_1 is the latent variable that represents Type A behaviour, and L_2 is the latent variable that represents autonomic arousal. The measured variables x_1 to x_5 are the items that assess Type A behaviour, the measured variables y_1 to y_4 are the items that assess autonomic arousal. The arrow labelled b is the effect of L_1 on L_2. The variance unexplained in L_2 is labelled u, and is equal to $1 - R^2$. It is LISREL convention to specify the predictor variables as the xs and outcome variables as the ys.

The ML estimation procedure estimates the path coefficients of the measurement model as before, but the procedure also estimates the path coefficient for the causal relationship between Type A behaviour and autonomic arousal.

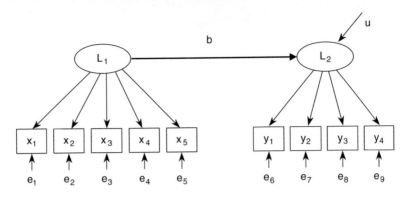

FIGURE 8.5 *A path model with two latent variables*

The correlation matrix shown in Table 8.5 represents the correlations between the four autonomic arousal variables (y_1 to y_4) and the five Type A behaviour measures (x_1 to x_5).

TABLE 8.5 *Correlation matrix of data on autonomic arousal and Type A behaviour;* N = 250

	y_1	y_2	y_3	y_4	x_1	x_2	x_3	x_4	x_5
y_1	1.00								
y_2	0.58	1.00							
y_3	0.39	0.45	1.00						
y_4	0.38	0.45	0.30	1.00					
x_1	0.26	0.31	0.20	0.20	1.00				
x_2	0.46	0.54	0.36	0.36	0.50	1.00			
x_3	0.23	0.27	0.18	0.18	0.25	0.44	1.00		
x_4	0.25	0.29	0.19	0.19	0.27	0.47	0.24	1.00	
x_5	0.44	0.51	0.34	0.34	0.47	0.83	0.42	0.45	1.00

When we analyse the model specified in the path diagram shown in Figure 8.5 we find that $\chi^2 = 35.8$, $df = 26$, $p = 0.09$. Because this result is not significant (>0.05), we can conclude that the model is an adequate representation of the data. We can also examine the path coefficients; the easiest way to do this is to put them into the path diagram. This path diagram is shown in Figure 8.6.

8.3 Programs for MLM and SEM

Research in psychology and its methodology tends to be updated rapidly, but textbooks can usually change fast enough to keep up. Software for carrying out statistical analysis changes even faster, so we are wary of giving information about programs when we know that by the time you read this, the

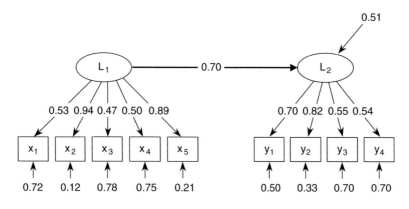

FIGURE 8.6 *Path diagram showing path coefficients*

information may well be out of date. Because of this, we recommend that you do not place a great deal of trust in this information, but instead refer to our sources of information that can be found on the World Wide Web pages associated with this book: these we can keep up to date. For accurate and up-to-date information, consult the Web pages of the publishers of the software. We intend to insert links from our pages to the publishers' pages. One point regarding SEM and MLM software is that, unlike larger general-purpose statistics packages, the programs tend to be written by a very small number (often one or two) of people. Every one of these authors is an active researcher with ideas and opinions of their own about how research should be carried out, and what areas are important, and they tend to be contributors to various email discussion lists about the subjects. Each program in some small way reflects the personality and interest of the person who wrote it.

We would also like to point out that this list should not be considered to be complete, but simply reflects programs that we are aware of, and familiar with, at the time of writing. Omission of any program should not be seen as expressing any form of disapproval of the program.

8.3.1 MLM software

8.3.1.1 MLn/MLWin

MLn and MLWin are written and published by Goldstein, Rasbash, and the Multilevel Modelling Group at the Institute of Education in London. Originally, the software was developed as ML2 (for the analysis of Level 2 models) and then ML3 (for Level 3) and then evolved into MLn (for Level n). The original versions were a command-line-based program with an interface that would be familiar to users of Minitab. The latest version at the time of writing is MLWin, which has a more user-friendly Windows interface.

MLn/MLWin is supported by Goldstein and the Multilevel Modelling Group. They produce an occasional newsletter for users of the program (and

MLM) and they run regular help sessions and workshops. Goldstein is also owner of the multilevel email list run by the UK service Mailbase (http://www.mailbase.ac.uk).

8.3.1.2 HLM
HLM is written by Bryk and Raudenbush, from the Longitudinal and Multi-level Modelling Project (LAMMP) in the School of Education, at Michigan State University; at the time of writing the most up-to-date version is 4.04. Bryk and Raudenbush have also published a book on MLM (see the section on further reading). HLM is a Windows-based program that allows you to specify the model parameters via menu commands.

8.3.1.3 MIXEDUP
MIXEDUP is a suite of four programs produced by Hedeker and Gibbons from the University of Illinois at Chicago for carrying out MLM. The four programs are MIXREG, for mixed regression models, with continuous data; MIXOR, for ordinal data; MIXPREG, for Poisson regression; and MIXNO, for nominal (logistic) regression. The programs all work in similar ways, and have an interface (in the Windows environment) that involves filling in 'cards'. The MIXEDUP suite of programs also has one final advantage: at the time of writing, it is free.

8.3.2 SEM software

8.3.2.1 LISREL
LISREL (Linear Structural Relationships) is written by Jöreskog and Sörbom, of Uppsala University in Sweden. It is published and supported by Scientific Software International, based in Chicago.

LISREL was the first commercially available program for SEM that we know of. Version 9.0 is promised for sometime in the near future. LISREL suffers from a perception problem: because it was the first program for SEM analysis it was designed for effectiveness rather than user-friendliness. In those days long ago when it was written, it was necessary to write the algorithms so that the program would run on the computers which then had limited capacity. The more recent versions of LISREL are updated versions and have user-friendly features similar to those in most other SEM programs, but it is remembered as being very difficult – if you hear people telling you this, be aware that their information may be out of date. LISREL can also carry out MLM, and exploratory factor analysis.

8.3.2.2 EQS
EQS (Equation Systems) is written by Bentler, of the University of California at Los Angeles, and published by Multivariate Software, Inc. EQS was a pioneering program arising out of first attempts to make SEM easier to

program. It used a language that is more intuitive and much easier to learn than the original LISREL language (although LISREL now has a similar language).

8.3.2.3 Mx

Mx, written by Neale from Virginia Commonwealth University, is unique amongst SEM programs in that it is free (it is even possible to run it over a Web page, and therefore not even necessary to possess a copy of the program).

Mx is a very general program in which the operations required must be specified in terms of the matrix algebra. Having to specify the operations in these terms makes it rather difficult to use unless the user is familiar with the workings of SEM, but what makes it difficult to use also makes it enormously flexible.

8.3.2.4 SEPath

SEPath is distributed with the general statistics package Statistica, and was written by Steiger, from the University of British Columbia. The user interface of SEPath is therefore much more closely integrated with a statistics package than other programs are. This integration with Statistica makes for a large bonus if you already happen to use Statistica. The unique selling point of SEPath is its ability to analyse a correlation matrix and provide appropriate standard errors and fit statistics.

8.3.2.5 AMOS

AMOS (Analysis of Moment Structures), written by Arbuckle, from Temple University, Philadelphia, is also currently available as an additional module in SPSS. Although most programs are capable of inputting a model as a path diagram, AMOS is unique in that the path diagram is not used to write syntax, which is then checked and run. Rather the path diagram acts as the syntax, and is run directly. AMOS also has powerful tools for bootstrap analyses and options for the analysis of missing data.

8.3.2.6 Mplus

Mplus, written and published by Muthen and Muthen, from University College, Los Angeles, is a relative newcomer to the SEM software fold, and we are not fully familiar with its capabilities. It is described as a 'second-generation' SEM program, and it is capable of, and goes some way towards, integrating a range of different methodologies, including SEM, MLM and latent class analysis.

Notes

1 Assuming the pigs can be kept still while they are weighed.
2 This may seem curious, that it is an advantage to be able to be wrong. In law, there is a right to be sued, which is also advantageous. If I am able to sue you, I am

more likely to lend you money, to buy a new computer. If you need a new computer (say, to finish a book on regression analysis) this is a useful right to have.

3 The alert reader may wonder why we have only four items to estimate, when the equation has four slopes and four error variances. We had to give the latent variable a variance, and we made this equal to one. The error variance must be equal to 1 − slope, so if we have estimated the slope, we have estimated the error also.

4 Other estimation procedures do exist, but ML is the most common.

Further reading

Because MLM and SEM are such complex topics, we provide below a fuller list of further reading, in contrast to other chapters.

Multilevel modelling
MLM is a relatively new technique and so few textbooks cover it. At the time of writing, we know of six books specifically about MLM.

Snijders and Bosker (1999), Hox (1995) and Kreft and De Leeuw (1998) are all good introductory texts for the beginner. (Hox is currently out of print, but is available as a free download from Hox's Web site.)

Bryk and Raudenbush (1992) provide a more detailed and elaborate treatment of the use of MLM, but they also make more demands on the reader's mathematical and statistical knowledge.

Goldstein (1995) and Longford (1993) are more advanced texts, and provide more mathematical details.

Structural equation modelling
SEM has been around a lot longer than MLM, so there are many more texts available which deal with it. We will mention below three types of text: introductory texts suitable for beginners, advanced texts, and texts which deal with specific issues.

Introductory texts
Loehlin (1998) *Latent variable models* (3rd ed.) is our favourite introductory text − it relates SEM to regression analysis and factor analysis, showing the clear links between them.

Schumacker and Lomax (1996) *A beginner's guide to structural equation modeling* is another excellent introductory text. It is a little more computer oriented than the other books for beginners, spending longer examining what actually to tell the computer and looking at output than, for example, Loehlin's text does.

Hayduk (1988) *Structural equation modeling with LISREL: essentials and advances* focuses mainly on LISREL, although it is actually a nice introduction to SEM. It focuses more on SEM than how to use LISREL.

Hoyle (1995) *Structural equation modeling: concepts, issues, and applications* is a very nice book, ideal for people just beyond the beginner stage in SEM. It contains chapters on the sorts of issues that people are likely to encounter once they start to use SEM 'in anger'. It also includes a very useful chapter on writing about structural equation models, and several chapters of examples of applications.

Advanced texts
Bollen (1989) *Structural equations with latent variables* is beginning to show its age a little, although it remains a classic. It is heavy going, but discusses issues such as identification and measurement error that would be difficult to find anywhere else.

Hayduk (1996) *LISREL: issues, debates, and strategies.* Not everyone will agree with everything Hayduk says – his opinions tend to be strongly held and not necessarily consensual. (See the January 2000 issue of the journal *Structural equation modelling* for more evidence of this.) This book is not for beginners, but raises interesting issues for more advanced students who want to know more about SEM.

Long (1983a and 1983b) *Confirmatory factor analysis: a preface to LISREL* and *Covariance structure models: an introduction to LISREL* are two books in the Sage University Series Quantitative Applications in the Social Sciences (or 'little green books', as they are more familiarly known). They are both showing their age a little, but are good, concise introductions to LISREL.

Specific topics
Bollen and Long (1993) *Testing structural equation models* is an amazingly useful book containing information on – well – testing structural equation models. In a range of chapters it describes not only how standard fit indices are calculated and what they mean, but also the interpretation of different fit indices and the general philosophy behind deciding whether a model 'fits'.

Jaccard and Wan (1996) *LISREL approaches to interaction effects in multiple regression* is another of the Sage 'little green books', this book focusing on methods of implementing interaction effects using LISREL. Like almost all of the little green books, it has a very process-driven approach to the subject, and is clearly written. Be warned, though, that the issue it deals with is not an easy one.

Marcoulides and Schumacker (1996) *Advanced structural equation modelling: issues and techniques.* The title tells you almost everything you need to know: this is an advanced text on SEM, so it is not for beginners. Its technical level is not completely impenetrable and a number of chapters are excellent sources of information on diverse topics.

Appendix 1 Equations

(If you really just want to see the equations, turn to the end of this appendix. We do, however, recommend that you read from the start.)

We have told you in this book that the regression equation represents the minimised sum of squared deviations from the regression line. We have also told you (in Chapter 6) that this is done using equations that are of a closed form. If you stick the numbers into the correct equation, solving the equation will give you all the information you need to know regarding the regression line. What we have not told you is what those equations are.

We have not told you because the equations, in our opinion, are not very important. What is much more important is a conceptual understanding of what those equations are trying to do.

Students sometimes say that they like to work through the equations, rather than do calculations on a computer, because they get a sense of what is happening – they complain when their tutors ask them to use computers to do the calculations for this reason. Well, we are sorry to be blunt, but we do not believe you. We think students do not like doing calculations on computers because it is not very easy. You have to think much more about what is going on. When we learned statistics computers had not been introduced to any great extent,[1] and we spent a lecture or seminar calculating ANOVA. We did not have to think about all of the tricky conceptual issues. We also do not think that the equations are sufficiently transparent to give students an idea of what is going on. The equations are algebraically manipulated to make them easier to do, with no regard for whether this makes it very difficult to see what is happening. (An example of this is the standard deviation calculation that we showed you in Chapter 1. We presented this equation so that you could see what was going on. However, this is not the equation you would use if you wanted to calculate the standard deviation; there is an easier formula, but it is not clear what the standard deviation actually is if you use this formula.) Our final point is that to work through some of the calculations we use in this book would take hours. A linear regression analysis with five independent variables would probably take someone with a calculator a couple of days, even if they were very familiar with the procedures. Then, if we wanted VIF or tolerance statistics we would have to do almost all of it again, for every single variable in our sample. And adding another independent variable does not just increase the work by a little, it more than doubles it.

So in this section we are going to present the equations to do regression analysis, but we are going to attempt to show not just what they do, but where possible *why* they do it.

A1.1 Calculating a covariance

Let us take a set of four numbers: 2, 4, 6, 8. The mean value of these four figures is:

$$\bar{x} = \frac{\sum x}{N} = \frac{2+4+6+8}{4} = \frac{20}{4} = 5$$

When estimating the correlation coefficient we are not interested in the mean, so we can subtract the mean from each score to create deviations from the mean. This is also called centring, and we used it in Chapter 7. The centred scores are shown in Table A1.1.

TABLE A1.1 Scores and centred scores

Subject	Original score (x)	Original score − mean (d)
A	2	−3
B	4	−1
C	6	1
D	8	3

We are now part of the way to calculating the variance. The formula for the variance is shown below, where d is the deviation of each score from the mean:

$$\text{Var}(x) = \frac{\sum d^2}{N}$$

$$\text{Var}(x) = \frac{(-3^2) + (-1^2) + (1^2) + (3^2)}{4}$$

$$= \frac{(9) + (1) + (1) + (9)}{4} = \frac{20}{4} = 5$$

In fact, this is the sample variance, which is a biased estimator of the population variance. To estimate the population variance (which is what we are really interested in) we need to divide by $N - 1$:[2]

$$\text{Var}(x) = \frac{\sum d^2}{N - 1}$$

When we substitute the values from Table A1.1 into this formula we get:

$$\text{Var}(x) = \frac{(-3^2) + (-1^2) + (1^2) + (3^2)}{3}$$

$$= \frac{9 + 1 + 1 + 9}{3} = \frac{20}{3} = 6.67$$

A different way of thinking about this would be to plot the variables on a scatter-graph, with the same variable on each axis. This is shown in Figure A1.1.

If we draw a line, perpendicular to each axis, to each of the points, we make a square for each point, as shown on the graph. The length of the side of each square is the distance between the value and the mean − it is the deviation from the mean. Because it is a square, the area that it encloses is equal to the length of the side, squared. Therefore, the total size of all four squares is equal to:

$$(-3^2) + (-1^2) + (1^2) + (3^2) = 9 + 1 + 1 + 9 = 20$$

218 APPLYING REGRESSION AND CORRELATION

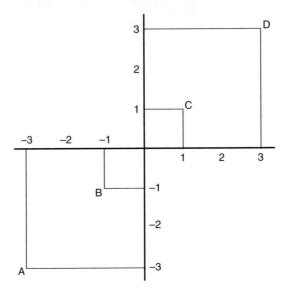

FIGURE A1.1

We can calculate the average size of the squares, by dividing this figure by N:

$$\frac{20}{4} = 5$$

What we have just done, in a roundabout sort of way, is to calculate the variance. Although again this is the sample variance, and we could divide by $N - 1$ to get the population estimate:

$$\frac{20}{3} = 6.67$$

What happens if instead of calculating the variance by plotting the same variable on both axes of the plot, we plot two variables, one on each axis? We will present another variable, to go with our original variable (x). This variable is called y. The variables x and y, and their respective deviations from the mean, are shown in Table A1.2.

TABLE A1.2

Subject	x	$x - \bar{x}$	y	$y - \bar{y}$
A	2	−3	4	−1
B	4	−1	2	−3
C	6	1	8	3
D	8	3	6	1

The variance of both x and y is 5, and the mean values of x and y are also 5. If we plot these variables together on a scattergraph, they will look as shown in Figure A1.2. We could also add rectangles to these, as shown in Figure A1.3.

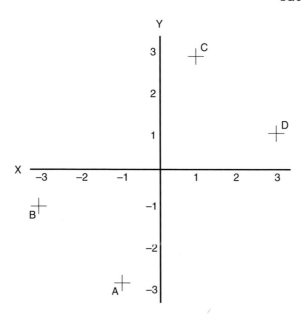

FIGURE A1.2 *Scatterplot of* x *and* y

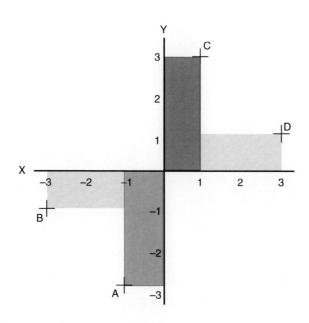

FIGURE A1.3

220 APPLYING REGRESSION AND CORRELATION

Now, instead of squares we have rectangles. The product of a rectangle's two sides gives its area. The average area of the rectangles is therefore equal to:

$$= \frac{(-1 \times -3) + (-3 \times -1) + (1 \times 3) + (3 \times 1)}{4} = \frac{3 + 3 + 3 + 3}{4} = \frac{12}{4} = 3$$

By now you probably realise that we are going to have to divide by $N - 1$:

$$\frac{12}{3} = 4$$

This value 4 is referred to as the *covariance* of x and y. It is the combined *variance* of the two variables. In this case its value of 4 is smaller than the variance of both of the variables. The covariance can *never* be higher than the variance of both of the variables (although it is possible for it to be higher than the variance of one of the variables).

The data for another example are shown in Table A1.3, and the scatterplot in Figure A1.4.

TABLE A1.3

Subject	x	$x - \bar{x}$	y	$y - \bar{y}$
A	2	-3	6	1
B	4	-1	8	3
C	6	1	2	-3
D	8	3	4	-1

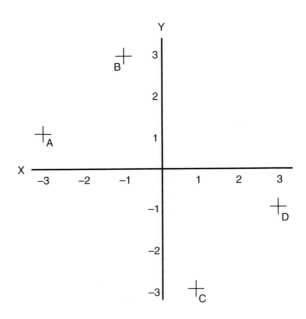

FIGURE A1.4

We will skip the step in which we drew the rectangles on the scatterplot, but calculate the areas of those rectangles. The average size of the rectangles is, as we have seen, called the covariance of x and y, which is written Cov(x,y):

$$\text{Cov}(x,y) = \frac{(-3 \times 1) + (-1 \times 3) + (1 \times -3) + (3 \times -1)}{4}$$

$$= \frac{(-3) + (-3) + (-3) + (-3)}{3} = \frac{-12}{3} = -4$$

Just like in correlations and in regression slopes, a positive covariance means that higher values of one variable are associated with higher values of another variable.

We can turn the procedure that we have just carried out into an equation:

$$\text{Cov}(x,y) = \frac{\Sigma(x - \bar{x})(y - \bar{y})}{N - 1}$$

A1.2 Calculating a correlation

A problem that we have with the covariance is that its magnitude is dependent not only upon the magnitude of the association between the two variables, but also upon the variances of the original variables. (You may remember that this is the same problem that we had in Chapter 1 with the regression coefficients.)

One way around this is to standardise the variables before we carry out the analysis. This is done in Chapter 1 by dividing each centred value by the standard deviation. We could then calculate the covariance. This is a little bit tricky, and involves more maths, which introduces the possibility for more mistakes. To get around this problem we will standardise the covariance *after* we have calculated it.

We standardise a variable by dividing it by its standard deviation. The standard deviation, you will recall, is the square root of the variance. We have calculated the covariance by multiplying variables together, not just one variable, as in the variance, so we will need to multiply the variances together. Then we need to take the square root of that product of the variances, and divide the covariance by that figure.

That all seems a bit complicated, but it leads to the (fairly) simple equation that the standardised covariance (r_{xy}) between x and y is:

$$r_{xy} = \frac{\text{Cov}(x, y)}{\sqrt{\text{Var}(x) \times \text{Var}(y)}}$$

We can write this out in full by substituting the equations that we know for each part of this equation, which would give us:

$$r_{xy} = \frac{\left(\frac{\Sigma(x - \bar{x})(y - \bar{y})}{N - 1}\right)}{\sqrt{\frac{\Sigma(x - \bar{x})^2}{N - 1} \times \frac{\Sigma(y - \bar{y})^2}{N - 1}}}$$

This type of standardised covariance is given a special name – it is called a correlation coefficient, written as r. The correlation between two variables x and y is written as r_{xy}. The equation seems very large, but it is made up, as we have seen, of a number of very simple components.

We can calculate the correlation between x and y, using the covariance, for the data in our first example:

$$r_{xy} = \frac{\mathrm{Cov}(x,y)}{\sqrt{\mathrm{Var}(x) \times \mathrm{Var}(y)}}$$

$$r_{xy} = \frac{4}{\sqrt{6.67 \times 6.67}}$$

$$= 0.6$$

This correlation, you will remember, is equal to multiple R when there is only one independent variable in a regression equation, and is also equal to standardised beta.

A1.3 Calculating a slope

At this point it starts to get a little more complicated, and proving some of the things we are going to say requires either a large amount of algebra, or some knowledge of calculus. We do not think that you want to read that sort of thing, so bear with us, and take our word for a couple of things.

The b coefficient can be calculated from the correlation. The correlation was a standardised covariance, where both of the variables were given a standard deviation equal to 1. We want to know how much the variable y changes, when x changes by one unit. The correlation tells us how many standard deviations (SDs) y changes when x increases by 1 SD. The first thing we need to do is to turn the units of y from SDs back to the real scaling. We do this simply by multiplying the correlation by the SD of y. This gives us the amount that y changes when x increases by 1 SD. But we are not interested in SDs of x. We are interested in real units of x. To find this out we have to divide by the SD of x. We have not yet calculated the SD of these variables, but remember that the SD is equal to the square root of the variance. Our value for b then becomes:

$$b = r_{xy} \times \frac{\mathrm{SD}(x)}{\mathrm{SD}(y)}$$

$$b = 0.6 \times \frac{\sqrt{6.67}}{\sqrt{6.67}}$$

$$= 0.6$$

You will notice that the slope is the same as the correlation; this will only occur if, as in our case, the two variances are the same.

A1.4 Calculating the intercept

We now have the slope, in real units, but this was calculated to go through zero (we centred both variables right at the start). We want to correct this value so that it is equal to the true constant, which is the point at which the line hits the y-axis.

The first thing to do is to add the mean of the y variable to the constant. When we do this the line will no longer hit the y-axis at zero, but will hit it at the mean value of y. But this value now gives the value of the intercept if the y-axis crossed the x-axis at the mean value of x, not at zero. So we need to correct for the mean of x, but unless the slope is equal to 1, we need to find out how much the line changes between the mean of x and the point at which $x = 0$. To do this we use the slope, because the change must be equal to the mean of x multiplied by the slope.

Therefore, the constant (c) is found using:

$$c = \bar{y} - b\bar{x}$$

Substituting the values from our example:

$$c = 5 - 5 \times 0.6 = 2.0$$

A1.5 Calculating slopes in multiple regression

We are sorry to admit that our powers of explanation are going to fail us in this section. There really is no way to explain this, other than using matrix algebra. We will very briefly explain a little about matrices, but we are not going to go into any depth. If you want to know more about how to do things with matrices, Hadi (1996) is a very straightforward, applied book that will take you through the details.

A1.5.1 Scalars, vectors and matrices

A scalar is a number. For example, the number 2 is a scalar. That is all.

A vector is a series of scalars, either a row or a column. A vector is called a column vector, or a row vector. The following is a column vector:

$$\begin{pmatrix} 2 \\ 3 \\ 5 \\ 1 \end{pmatrix}$$

whereas:

$$(2\ 5\ 9\ 1\ 1\ 2\ 5)$$

is a row vector. Every time we have talked about a variable, we could have talked about a vector. When describing a vector, we put the number of rows first, followed by the number of columns. The first example was a 4×1 vector. The second example was a 1×7 vector. A scalar (or number) is a 1×1 vector.

Matrices are like a series of vectors. They have more than one row and more than one column. The following is a 4×5 matrix:

$$\begin{pmatrix} 2 & 3 & 7 & 6 & 9 \\ 3 & 1 & 6 & 8 & 3 \\ 7 & 7 & 0 & 4 & 3 \\ 4 & 6 & 0 & 6 & 5 \end{pmatrix}$$

Correlation matrices (and also covariance matrices) are a special type of matrix, called a square, symmetrical matrix. They are square because they have an equal number of rows and columns, and they are symmetrical because the top right half is equal to the bottom left half.

The following matrix is the correlation matrix of the variables books, attend, bxa and grade from Chapter 7:

$$\begin{pmatrix} 1.000 & 0.444 & -0.058 & 0.492 \\ 0.444 & 1.000 & -0.063 & 0.482 \\ -0.058 & -0.063 & 1.000 & 0.230 \\ 0.492 & 0.482 & 0.230 & 1.000 \end{pmatrix}$$

The series of ones, going from top left to bottom right is called the main diagonal. Correlation matrices always have ones in the main diagonal.

There is a special type of matrix called the identity matrix. This is a square matrix which has ones in the diagonal, and zeros in all of the other values. This is a 3 × 3 identity matrix:

$$\begin{pmatrix} 1 & 0 & 0 \\ 0 & 1 & 0 \\ 0 & 0 & 1 \end{pmatrix}$$

The identity matrix is given the symbol **I**. The number of rows and columns is indicated by a subscript. The previous matrix would be called I_3. I_1 is the number (or scalar) 1.

Matrices can be multiplied by scalars, vectors, or matrices. Multiplying by a scalar is easy – just multiply all of the elements of the matrix by the scalar. Multiplying by vectors is less easy. The following shows how to multiply a 3 × 3 matrix by a 3 × 1 vector. The result is a 3 × 1 vector. The labels in the vector show how the result was calculated. The first value in the vector is equal to $a \times j + d \times k + g \times l$, and so on.

$$\begin{pmatrix} a & d & g \\ b & e & h \\ c & f & i \end{pmatrix} \times \begin{pmatrix} j \\ k \\ l \end{pmatrix} = \begin{pmatrix} aj + dk + gl \\ bj + ek + hi \\ cj + fk + il \end{pmatrix}$$

Matrices can also be multiplied by other matrices. Using the same system as previously:

$$\begin{pmatrix} a & b \\ c & d \end{pmatrix} \times \begin{pmatrix} e & f \\ g & h \end{pmatrix} = \begin{pmatrix} ae + cf & af + bh \\ ce + dg & cf + dh \end{pmatrix}$$

Something important to note about the multiplication of vectors and matrices is that they are not commutative. This means that, if **A** and **B** are matrices, multiplying **A** × **B** is not the same as multiplying **B** × **A**.

The identity matrix can be thought of as being similar to the number 1. (In fact the scalar (1) is a 1 × 1 identity matrix.) Multiplying a matrix by the identity matrix (**I**) has no effect on the matrix, in the same way that multiplying a number by 1 has no effect. So:

$$\begin{pmatrix} 1.000 & 0.444 & -0.058 & 0.492 \\ 0.444 & 1.000 & -0.063 & 0.482 \\ -0.058 & -0.063 & 1.000 & 0.230 \\ 0.492 & 0.482 & 0.230 & 1.000 \end{pmatrix} \times \begin{pmatrix} 1 & 0 & 0 & 0 \\ 0 & 1 & 0 & 0 \\ 0 & 0 & 1 & 0 \\ 0 & 0 & 0 & 1 \end{pmatrix} = \begin{pmatrix} 1.000 & 0.444 & -0.058 & 0.492 \\ 0.444 & 1.000 & -0.063 & 0.482 \\ -0.058 & -0.063 & 1.000 & 0.230 \\ 0.492 & 0.482 & 0.230 & 1.000 \end{pmatrix}$$

Numbers have an inverse. The inverse of 5 is 1/5, the inverse of 20 is 1/20. If we multiply a number by its inverse, we are left with the number 1:

$$5 \times \frac{1}{5} = 1$$

$$20 \times \frac{1}{20} = 1$$

In much the same way, many (but not all) matrices have an inverse. If we multiply a matrix by its inverse, we are left with the identity matrix. The inverse of the matrix $\mathbf{A}$ is called $\mathbf{A}^{-1}$. Finding the inverse of a matrix is astonishingly hard work if your matrix is larger than 2 × 2 (and even then it is not very easy). Inverting matrices is what made regression analysis so difficult in the days before computers. To invert a 2 × 2 matrix is not very difficult. To invert a 3 × 3 matrix, we must invert four 2 × 2 matrices. To invert a 4 × 4 matrix, we must invert four 3 × 3 matrices (each of which involves inverting four 2 × 2 matrices). It rapidly becomes clear why people thought long and hard before they added another independent variable to a regression equation.

However, not all matrices can be inverted. A correlation matrix in which any two variables correlate to 1.00 will not have an inverse. Similarly, if the multiple correlation between any independent variable and any combination of other independent variables is equal to 1.00 the matrix will not have an inverse. This is why regression analysis makes an assumption that there will not be complete collinearity. If there is, the matrix cannot be inverted, and as we shall see in a moment, we have to invert a matrix to do regression. Multiplying and inverting matrices is fairly straightforward in a lot of computer packages, including statistics packages like SPSS and Minitab, most modern spreadsheet packages (including Microsoft Excel, Lotus 1–2–3, Borland Quattro Pro), and specialised mathematics programs (e.g. MathCad and Mathematica).

To calculate the standardised b coefficients, we multiply the inverse of the matrix of correlations between the independent variables by the vector of correlations between the dependent variables. The first step is to find the inverse of the matrix of correlations between the independent variables (you could work this out for yourself using a computer package, or simply believe us):

$$\begin{pmatrix} 1.000 & 0.444 & -0.058 \\ 0.444 & 1.000 & -0.063 \\ -0.058 & -0.063 & 1.000 \end{pmatrix}^{-1} = \begin{pmatrix} 1.246 & -0.551 & 0.037 \\ -0.551 & 1.247 & 0.047 \\ 0.037 & 0.047 & 1.005 \end{pmatrix}$$

The next step is to multiply this matrix by the vector of correlations between the independent variables and dependent variable:

$$\begin{pmatrix} 1.246 & -0.551 & 0.037 \\ -0.551 & 1.247 & 0.047 \\ 0.037 & 0.047 & 1.005 \end{pmatrix} \times \begin{pmatrix} 0.492 \\ 0.482 \\ 0.230 \end{pmatrix} = \begin{pmatrix} 0.356 \\ 0.341 \\ 0.272 \end{pmatrix}$$

The resulting 3 × 1 vector shows the standardised b coefficients given by the regression output. We could use the standard deviations of the variables to calculate the unstandardised slopes, and then the method displayed previously to calculate the intercept.

The final thing to point out, before we move on, is that if the independent variables are uncorrelated, the correlation matrix of the independent variables will be an identity matrix. The inverse of an identity matrix is also an identity matrix, so we would multiply an identity matrix by the vector of correlations between the independent variables and the dependent variable. Because multiplying a vector by an identity matrix leaves the vector unchanged, the standardised beta weights are the correlations.

A1.6   Calculating multiple R

We saw in Chapter 1 that we could calculate multiple R as the correlation between the estimated values of the dependent variable and the actual values of the dependent variables.

When we have calculated the unstandardised slopes and the constant, we can use the equation to get predicted values for the dependent variable, and calculate R.

A1.7   Just the equations

Covariance of x and y:

$$\text{Cov}(x,y) = \frac{\sum(x - \bar{x})(y - \bar{y})}{N - 1}$$

Correlation of x and y:

$$\frac{\text{Cov}(x, y)}{\sqrt{\text{Var}(x) \times \text{Var}(y)}}$$

or:

$$r_{xy} = \frac{\left(\frac{\sum(x - \bar{x})(y - \bar{y})}{N - 1}\right)}{\sqrt{\frac{\sum(x - \bar{x})^2}{N - 1} \times \frac{\sum(y - \bar{y})^2}{N - 1}}}$$

Unstandardised slope, from correlation or standardised slope:

$$b = r_{xy} \times \frac{\text{SD}(x)}{\text{SD}(y)}$$

Intercept:

$$c = \bar{y} - b\bar{x}$$

Standardised slopes in multiple regression:

Inverse of matrix of correlations of IVs × vector of correlations of IVs with DV

Notes

1 We now sound old and boring.

2 This is all because of degrees of freedom (dfs). We can estimate N parameters from a dataset (where N is the sample size) and at this point we will have completely explained the data. We have already estimated one parameter, the mean, and this has 'used up' one df. We are therefore only left with four dfs to estimate the SD. (We would have three to estimate another parameter, such as kurtosis.)

Appendix 2 Doing regression with SPSS

In this appendix we will look at regression analysis and a few other little things in SPSS for Windows™ version 9.0. Instructions for some other programs are available on the Web site related to the book (http://www.jeremymiles.co.uk/regressionbook).

A2.1 SPSS

In this section we will describe how SPSS for Windows version 9.0 may be used to carry out some of the procedures we have described in the book. The regression procedure in SPSS has changed only a little since the Windows version was released, so at present you should be able to follow everything if you are using any version from 5.0 for Windows onwards. Of course we do not know what will happen, but we have no reason to believe that SPSS will change dramatically in the future.

A2.1.1 Linear regression

A2.1.1.1 Main dialog box

To find the main regression dialog box, click on the **Analyze** menu item (labelled **Statistics** in older versions of SPSS). Then click on **Regression**, and then **Linear** . . . You should see something like Figure A2.1. We have given the labels A to M to the different parts of the window, which are referred to below. Anything we have not bothered to label is not mentioned anywhere else in the book.

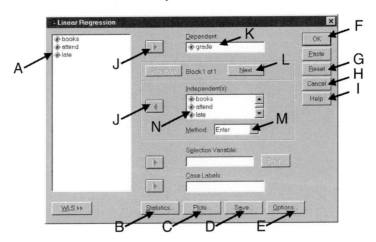

FIGURE A2.1 *Linear regression dialog box in SPSS*

DOING REGRESSION WITH SPSS

A: The list of dependent variables in your dataset. Depending on how SPSS is set up, you might see the variable labels here, rather than the variable names.
B: Choose some additional statistics to go in the output. See the section below.
C: Choose what plots you would like drawn. See below.
D: Choose what information you would like to save in your data file. See below.
E: Some additional options. See below.
F: **OK**. Press this when you have finished, to run the analysis.
G: Reset all values back to their defaults. Useful if you want to run a completely different analysis.
H: Cancel and ignore any changes that have been made.
I: **Help**. Get some help.
J: These buttons move variables between the variable list on the left (A), and the independent and dependent boxes (N and K).
K: The dependent variable. Use the button (J) to put your dependent variable into this box.
L: Next block. This is used when carrying out hierarchical regression (see Chapter 2) to add variables in blocks.
M: The variable selection technique. Choices are **Enter, Stepwise, Remove, Backward, Forward**. See Chapter 2.
N: The list of dependent variables.

A2.1.1.2 Statistics dialog box

Figure A2.2 shows the dialog box that appears when you click the save button.

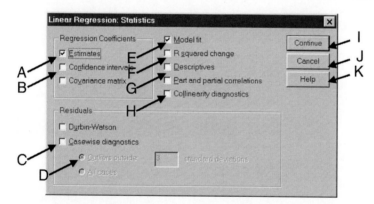

FIGURE A2.2 *Statistics dialog box*

A: Estimates. Tick this box to get the parameter estimates (the constant and slope coefficient for the variables). It is ticked by default.
B: Confidence intervals. Tick this box to get confidence intervals around the parameter estimates. These are calculated from the standard errors.
C: Casewise diagnostics of the residuals. This will print out details of any outliers which are further from the mean than you specify in D. See Chapter 4.
D: Tells SPSS which outliers to print out.
E: Model fit. Tick this box to get R, R^2, adjusted R^2 and the ANOVA table. It is ticked by default.
F: R^2 change. Tick this box to get a value, and significance test for the change in R^2 when additional blocks are entered. This is used when carrying out hierarchical regression (see Chapter 2).
G: Descriptive statistics for all of your variables – means, standard deviations, etc.
H: Tick this box to get the tolerance and VIF (Chapter 6).

I: **Continue.** Click here when finished.
J: **Cancel.** Go back and ignore all changes.
K: **Help.** Get additional help.

A2.1.1.3 Plots dialog box
This is shown in Figure A2.3.

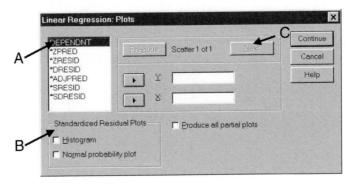

FIGURE A2.3 *Plots dialog box*

A: The list of possible variables to plot. (See Chapter 4 for descriptions.)
B: Residual plots, to check for normality.
C: Next button, to add more plots.

A2.1.1.4 Save dialog box
This dialog box (Figure A2.4) allows you to calculate and save a range of variables in your dataset.

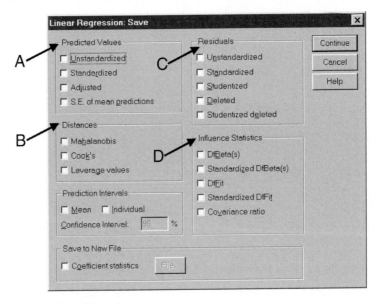

FIGURE A2.4 *Save dialog box*

A: Predicted values (see Chapters 1, 2 and 4). Adjusted values are the predicted value for the case when the analysis was carried out without the case.
B: Distance statistics. See Chapter 4 for descriptions.
C: Residuals. See Chapters 1 and 4 for descriptions.
D: Influence statistics (see Chapter 4).

A2.1.1.5 Options dialog box
This is shown in Figure A2.5.

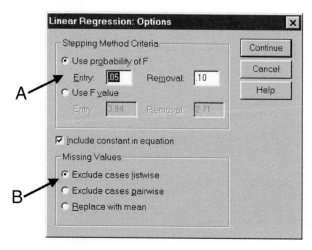

FIGURE A2.5 *Options dialog box*

A: Alter the entry and exit criteria for stepwise methods.
B: Determine what to do with missing data.

A2.1.2 Computing and recoding variables

There are two ways in SPSS of calculating new variables from old. When you are dealing with continuous data, the most convenient method is the compute dialog. When dealing with categorical variables, for example if you want to create dummy variables (see Chapter 3), the most convenient method is the recode dialog.

A2.1.2.1 The compute dialog

When we want to carry out some sort of transformation on a scale, for example a non-linear transformation (see Chapter 6), or if we want to create a moderator (see Chapter 7) we can use the compute variable dialog (Figure A2.6). This is found in the data window, by clicking on **Transform** menu, followed by **Compute**.

A: Target variable. This is the name of the new variable to be created.
B: Variable list. The name of the variables in the data file.
C: Numeric expression. Here we put the calculation that will be used to create the new variable.
D: Some mathematical functions to put in the numeric expression.
E: Functions to add in the numeric expression, for example log (see Chapter 6).

Some examples of the compute box:

232 APPLYING REGRESSION AND CORRELATION

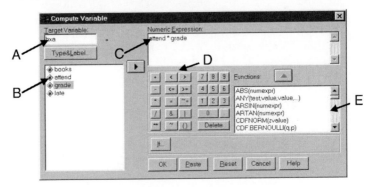

FIGURE A2.6 *Compute variable dialog box*

To create a new variable called bxa, which is the product of books and attend:

- In the target variable box, write bxa.
- In the numeric expression write books * attend.
- Press **OK**.

To create a new variable called hassles3, which is the cube of hassles:

- In the target variable box, write hassles3.
- In the numeric expression write hassles ** 3.
- Press **OK**.

To create a new variable called logtime, which is the log of a variable called time:

- In the target variable box, write logtime.
- In the numeric expression write log(time).
- Press **OK**.

A2.1.2.2 The recode dialog box

The recode dialog box is used when we want to manipulate categorical variables. This is most commonly done when we want to take a categorical variable with more than two levels, and turn it into a series of dummy-coded variables (see Chapter 3).

Imagine we have a categorical variable, called group, which has three possible values. The value 0 indicates that the person is in the Control Group, 1 indicates they are in Group 1, and 2 indicates they are in Group 2. We want to turn the variable group into two dummy-coded variables that represent membership of Group 1 and Group 2. (We do not need a third variable to represent membership of the control groups – see Chapter 3 for an explanation of why.) We will call the two new dummy variables group_1, which will be equal to one if the person is in Group 1, and zero otherwise, and group_2, which will be equal to one if the person is in Group 2, and zero otherwise. Table A2.1 shows the possible values of the three variables.

TABLE A2.1

group	group_1	group_2
0	0	0
1	1	0
2	0	1

DOING REGRESSION WITH SPSS   233

The recode process has two steps. In the first step we tell SPSS what name we would like the new variable to have. In the second step we tell SPSS what we would like the values in the new variable to be. To recode variables, first select the **Transform** menu, then the **Recode** item, and then **Into Different Variables** . . .

Step 1: Name new variables. The dialogue box to do this is shown in Figure A2.7.

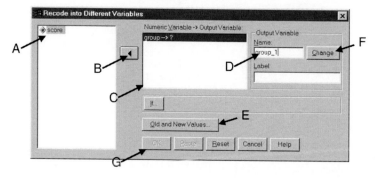

FIGURE A2.7

A: The variable list, with which we are becoming familiar.
B: The button to move variables to and from the variable list.
C: The list of old variables, linked to their new variables.
D: The list of new variables. Type the name of the new variable here, and then click on the change button (F). In our example we would first type group_1 here.
E: Use this button to select the next dialog box, where we tell SPSS what values are to change.
F: The change button adds the name of the variable that we have typed into D to the output variable list in C.
G: The OK button. You will not be able to press this until you have pressed button E, and set old and new values.

Step 2: Tell SPSS the old and new values. The dialog box to do this is shown in Figure A2.8.

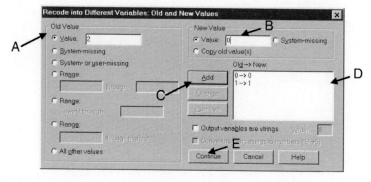

FIGURE A2.8

A: Here we insert the value of the variable we want to recode.
B: Here we insert the value we want to have in the new variable.

234 APPLYING REGRESSION AND CORRELATION

C: When we have filled in both A and B, we click the add button to add the recode to the list.
D: Here we have the list of transformations that are to take place. In our example, when we are coding the variable `group_1`, this should say:

- 0–>0
- 1–>1
- 2–>0

When we are recoding the variable `group_2`, this should say:

- 0–>0
- 1–>0
- 2–>1

Note that we need to do two 'runs' through this procedure to create both of the dummy variables.

E: **Continue**. Click on this when you have finished.

A2.1.3 Logistic regression

The logistic regression procedure is more sophisticated than the linear regression procedure, and automates some of the tasks that we had to do to prepare data for categorical independent variables (see Chapter 3) and interactions (see Chapter 6).

To select the logistic regression dialog, choose the **Analyze** menu (**Statistics** in some versions of SPSS). Then choose the **Regression** item, and then **Binary Logistic** (just **Logistic** in some versions).

A2.1.3.1 The logistic regression dialog box
This is shown in Figure A2.9.

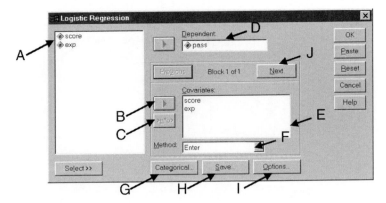

FIGURE A2.9 *Logistic regression dialog box*

A: The variable list.
B: The button to move variables to and from the variable list.

C: This is an interaction button. It allows us to specify an interaction effect from this dialog box without going through the steps we described in Chapter 7. To use this, select the two variables you are interested in, and then instead of using the usual button to move them, use this button.
D: The dependent variable. This must have only two values.
E: The covariates list. This is the word that SPSS uses for independent variables in logistic regression.
F: The entry method. See Chapter 2 for a description of these methods. (And we emphasise everything we said about stepwise methods for linear regression models in Chapter 2.)
G: Here we can specify that variables are categorical. This saves us going through the procedure described in Chapter 3 to create dummy variables (see below).
H: The save button. This allows us to save some information back to the dataset (see below).
I: Options: See below.

A2.1.3.2 The categorical variable dialog box
This is shown in Figure 2.10.

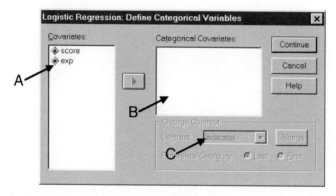

FIGURE A2.10 *Categorical variable dialog box*

A: The variable list.
B: The list of categorical variables. If you have any categorical independent variables, you can add them to this list. This saves going through the procedure that we described in Chapter 3 to create dummy variables.
C: Here the type of coding is specified. These were described in Chapter 3. First you select the reference category (first or last), and then the type of coding you would like to use. The two main choices are indicator coding and what SPSS calls simple coding, which we described as dummy coding in Chapter 3.

A2.1.3.3 Save dialog box
The options in this box (Figure A2.11) are very similar to the options that were available in the save dialog box in linear regression. Most of the diagnostic checks that are available in logistic regression are very similar to the checks that were available in linear regression, which we dealt with in Chapter 4. We did not consider these options specifically in terms of linear regression, though, and we suggest that you see a more comprehensive text on logistic regression, for example Menard (1995).

236 APPLYING REGRESSION AND CORRELATION

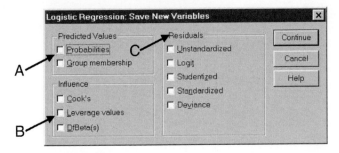

FIGURE A2.11 *Save dialog box*

A: Predicted values. These are either in terms of probability of group membership, or in terms of predicted group membership, and are based on a cut-off of 0.5.
B: Influence statistics. These are the equivalent of the influence statistics we encountered for linear regression in Chapter 4.
C: Residuals. Again these are similar to the residuals we encountered in Chapter 4.

A2.1.3.4 Options dialog box
This is shown in Figure A2.12.

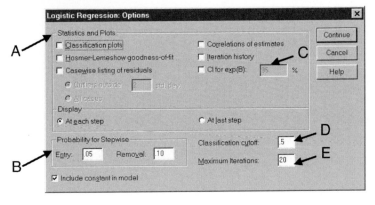

FIGURE A2.12 *Options dialog box*

A: A series of diagnostic charts and indices are presented within this section. Most of them are beyond the scope of this book.
B: The probabilities for stepwise entry and removal.
C: The 95% CI (confidence interval) for B. This is calculated using the standard error of B, as in linear regression (Chapter 1).
D: The classification cut-off. This is used to determine predicted group membership. If the probability of a person being in a group is greater than 0.5, the analysis predicts that the person will be in that group.
E: The maximum number of iterations that are allowed. On some occasions you may find that the logistic regression does not converge, and therefore you may need to increase this value.

Appendix 3 Statistical tables

This appendix contains some statistical tables that we have referred to in the book. We do not include every table that you might want to use, and we have not put as much detail in as we would like to. We do not recommend that you rely on these tables, rather we suggest you use them as a guide. We present fuller versions of these tables on the web pages associated with this book.

TABLE C.1 is used to help you determine whether a particular score is an outlier, based on its z-score, and assuming a normal distribution. The column labelled 'Proportion' provides an estimate of the proportion of people that should achieve a particular score.

Example: We have a measure, which is normally distributed, and we have a sample of size 100. We can see that a z-score of 1 would be achieved by one person in three, if the data were normally distributed. In this case a z-score of 1 should not cause us any concerns. A z-score of 2 should be achieved by approximately one person in 22, and again should not worry us. A z-score of 3 should be achieved by approximately one person in 370, and therefore may begin to make us want to examine the data more closely, but should still not be anything to worry about – 1 dataset in 4, of sample size 100 will contain such a measure. However, a z-score of 4 should only be achieved by approximately one person in 15 780, and therefore if any of the participants in our study have a z-score this high, we would probably want to investigate this case further.

TABLE C.2 If you are familiar with other statistical tables of critical values for r, or other statistics you may find this table a little confusing, but bear with us, and we will explain it. To use the table, when you have the value for r, find that value in the left-hand column. If the exact figure is not there, then use the next lowest one down. Then in the top row, find the value for your sample size (N). If the exact figure is not there, then again, use the next lowest one down. Where the row and column meet, you will find the probability value associated with those values of r and N.

Example: $r = 0.363$, $N = 47$. Find value next lowest to 0.363 in r column: 0.36. Find value below 47 in df row = 45. Read off probability: 0.015. If the exact values for r and N were not available in the table, you should be aware that your probability will be slightly lower than this.

Note that these probability values are all 2-tailed. For one-tailed probabilities, you should halve the values given.

$$*= <0.0005.$$

TABLE C.3 If you are familiar with other statistical tables of critical values for t, or other statistics you may find this table a little confusing, but bear with us, and we will explain it. To use the table, when you have the value for t, find that value in the left hand column. If the exact figure is not there, then use the next lowest one down. Then in the top row, find the value for your df. If the exact figure is not there, then again, use the next lowest one down. Where the row and column meet, you will find the probability value associated with those values of t and df.

Example: $t = 2.18$, df $=3$ 3. Find value next lowest to 2.18 in t column: 2.1. Find value below 33 in df row = 30. Read off probability: 0.044. If the exact values for t and df were not available in the table, you should be aware that your probability will be slightly lower than this.

TABLE C.1 *Normal distribution*

z-score	Proportion
1	3
1.1	4
1.2	4
1.3	5
1.4	6
1.5	7
1.6	9
1.7	11
1.8	14
1.9	17
2	22
2.1	28
2.2	36
2.3	47
2.4	61
2.5	81
2.6	107
2.7	144
2.8	196
2.9	268
3	370
3.1	517
3.2	728
3.3	1 034
3.4	1 484
3.5	2 149
3.6	3 142
3.7	4 637
3.8	6 909
3.9	10 392
4	15 780

Note that these probability values are all 2-tailed. For one-tailed probabilities, you should halve the values given.

TABLE C.4 If you are familiar with other statistical tables of critical values for χ^2, or other statistics you may find this table a little confusing, but bear with us, and we will explain it. To use the table, when you have value for χ^2, find that value in the left hand column. If the exact figure is not there, then use the next lowest one down. Then in the top row, find the value for your df. Where the row and column meet, you will find the probability value associated with those values of χ^2 and df.

Example: $\chi^2 = 6.44$, df $= 3$. Find value next lowest value to 6.44 in χ^2 column: 6.4. Find value 3 in df. Read off probability: 0.094.

TABLE C.5 Unfortunately space does not permit us to present the tables for the F distribution in the same detail as the previous distributions – complete F tables can go on for pages. This table shows only the critical value for the 0.05 probability. If your value of F exceeds the value stated in the table, then the probability of that value of F occurring is equal to, or less than, 0.05. If you require more accurate tables, you can consult the web pages associated with the book, or alternatively most spreadsheets, and maths or statistics packages contain functions to calculate probabilities of the F distribution.

TABLE C.2 P values for correlations

	N													
	10	15	20	25	30	35	40	45	50	60	70	80	90	100	200	500
0.1	0.783	0.723	0.675	0.634	0.599	0.568	0.539	0.513	0.490	0.447	0.410	0.377	0.348	0.322	0.159	0.025
0.11	0.762	0.696	0.644	0.601	0.563	0.529	0.499	0.472	0.447	0.403	0.365	0.331	0.302	0.276	0.121	0.014
0.12	0.741	0.670	0.614	0.568	0.528	0.492	0.461	0.432	0.406	0.361	0.322	0.289	0.260	0.234	0.091	0.007
0.13	0.720	0.644	0.585	0.536	0.494	0.457	0.424	0.395	0.368	0.322	0.283	0.250	0.222	0.197	0.067	0.004
0.14	0.700	0.619	0.556	0.504	0.461	0.422	0.389	0.359	0.332	0.286	0.248	0.215	0.188	0.165	0.048	0.002
0.15	0.679	0.594	0.528	0.474	0.429	0.390	0.356	0.325	0.298	0.253	0.215	0.184	0.158	0.136	0.034	0.001
0.16	0.659	0.569	0.500	0.445	0.398	0.359	0.324	0.294	0.267	0.222	0.186	0.156	0.132	0.112	0.024	*
0.17	0.639	0.545	0.474	0.417	0.369	0.329	0.294	0.264	0.238	0.194	0.159	0.132	0.109	0.091	0.016	*
0.18	0.619	0.521	0.448	0.389	0.341	0.301	0.266	0.237	0.211	0.169	0.136	0.110	0.090	0.073	0.011	*
0.19	0.599	0.498	0.422	0.363	0.315	0.274	0.240	0.211	0.186	0.146	0.115	0.091	0.073	0.058	0.007	*
0.2	0.580	0.475	0.398	0.338	0.289	0.249	0.216	0.188	0.164	0.125	0.097	0.075	0.059	0.046	0.005	*
0.21	0.560	0.453	0.374	0.314	0.265	0.226	0.193	0.166	0.143	0.107	0.081	0.062	0.047	0.036	0.003	*
0.22	0.541	0.431	0.351	0.291	0.243	0.204	0.173	0.146	0.125	0.091	0.067	0.050	0.037	0.028	0.002	*
0.23	0.523	0.410	0.329	0.269	0.221	0.184	0.153	0.129	0.108	0.077	0.055	0.040	0.029	0.021	0.001	*
0.24	0.504	0.389	0.308	0.248	0.201	0.165	0.136	0.112	0.093	0.065	0.045	0.032	0.023	0.016	0.001	*
0.25	0.486	0.369	0.288	0.228	0.183	0.147	0.120	0.098	0.080	0.054	0.037	0.025	0.017	0.012	*	*
0.26	0.468	0.349	0.268	0.209	0.165	0.131	0.105	0.085	0.068	0.045	0.030	0.020	0.013	0.009	*	*
0.27	0.451	0.330	0.250	0.192	0.149	0.117	0.092	0.073	0.058	0.037	0.024	0.015	0.010	0.007	*	*
0.28	0.433	0.312	0.232	0.175	0.134	0.103	0.080	0.062	0.049	0.030	0.019	0.012	0.008	0.005	*	*
0.29	0.416	0.294	0.215	0.160	0.120	0.091	0.069	0.053	0.041	0.025	0.015	0.009	0.006	0.003	*	*
0.3	0.400	0.277	0.199	0.145	0.107	0.080	0.060	0.045	0.034	0.020	0.012	0.007	0.004	0.002	*	*
0.31	0.383	0.261	0.183	0.132	0.095	0.070	0.052	0.038	0.028	0.016	0.009	0.005	0.003	0.002	*	*
0.32	0.367	0.245	0.169	0.119	0.085	0.061	0.044	0.032	0.023	0.013	0.007	0.004	0.002	0.001	*	*
0.33	0.352	0.230	0.155	0.107	0.075	0.053	0.038	0.027	0.019	0.010	0.005	0.003	0.001	0.001	*	*
0.34	0.336	0.215	0.142	0.096	0.066	0.046	0.032	0.022	0.016	0.008	0.004	0.002	0.001	0.001	*	*
0.35	0.321	0.201	0.130	0.086	0.058	0.039	0.027	0.018	0.013	0.006	0.003	0.001	0.001	*	*	*
0.36	0.307	0.187	0.119	0.077	0.051	0.034	0.023	0.015	0.010	0.005	0.002	0.001	*	*	*	*

continued overleaf

TABLE C.2 (cont.)

	N															
	10	15	20	25	30	35	40	45	50	60	70	80	90	100	200	500
0.37	0.293	0.175	0.108	0.069	0.044	0.029	0.019	0.012	0.008	0.004	0.002	0.001	*	*	*	*
0.38	0.279	0.162	0.098	0.061	0.038	0.024	0.016	0.010	0.006	0.003	0.001	0.001	*	*	*	*
0.39	0.265	0.151	0.089	0.054	0.033	0.021	0.013	0.008	0.005	0.002	0.001	*	*	*	*	*
0.4	0.252	0.140	0.081	0.048	0.029	0.017	0.011	0.006	0.004	0.002	0.001	*	*	*	*	*
0.41	0.239	0.129	0.073	0.042	0.024	0.014	0.009	0.005	0.003	0.001	*	*	*	*	*	*
0.42	0.227	0.119	0.065	0.037	0.021	0.012	0.007	0.004	0.002	0.001	*	*	*	*	*	*
0.43	0.215	0.110	0.058	0.032	0.018	0.010	0.006	0.003	0.002	0.001	*	*	*	*	*	*
0.44	0.203	0.101	0.052	0.028	0.015	0.008	0.004	0.002	0.002	0.001	*	*	*	*	*	*
0.45	0.192	0.092	0.046	0.024	0.013	0.007	0.004	0.002	0.001	*	*	*	*	*	*	*
0.46	0.181	0.084	0.041	0.021	0.011	0.005	0.003	0.001	0.001	*	*	*	*	*	*	*
0.47	0.170	0.077	0.037	0.018	0.009	0.004	0.002	0.001	0.001	*	*	*	*	*	*	*
0.48	0.160	0.070	0.032	0.015	0.007	0.004	0.002	0.001	*	*	*	*	*	*	*	*
0.49	0.151	0.064	0.028	0.013	0.006	0.003	0.001	0.001	*	*	*	*	*	*	*	*
0.5	0.141	0.058	0.025	0.011	0.005	0.002	0.001	*	*	*	*	*	*	*	*	*
0.51	0.132	0.052	0.022	0.009	0.004	0.002	0.001	*	*	*	*	*	*	*	*	*
0.52	0.123	0.047	0.019	0.008	0.003	0.001	0.001	*	*	*	*	*	*	*	*	*
0.53	0.115	0.042	0.016	0.006	0.003	0.001	*	*	*	*	*	*	*	*	*	*
0.54	0.107	0.038	0.014	0.005	0.002	0.001	*	*	*	*	*	*	*	*	*	*
0.55	0.100	0.034	0.012	0.004	0.002	0.001	*	*	*	*	*	*	*	*	*	*
0.56	0.092	0.030	0.010	0.004	0.001	*	*	*	*	*	*	*	*	*	*	*
0.57	0.085	0.027	0.009	0.003	0.001	*	*	*	*	*	*	*	*	*	*	*
0.58	0.079	0.023	0.007	0.002	0.001	*	*	*	*	*	*	*	*	*	*	*
0.59	0.073	0.021	0.006	0.002	0.001	*	*	*	*	*	*	*	*	*	*	*
0.6	0.067	0.018	0.005	0.002	*	*	*	*	*	*	*	*	*	*	*	*
0.61	0.061	0.016	0.004	0.001	*	*	*	*	*	*	*	*	*	*	*	*
0.62	0.056	0.014	0.004	0.001	*	*	*	*	*	*	*	*	*	*	*	*

0.63	0.051	0.012	0.003	0.001
0.64	0.046	0.010	0.002	0.001
0.65	0.042	0.009	0.002	*
0.66	0.038	0.007	0.002	*
0.67	0.034	0.006	0.001	*
0.68	0.031	0.005	0.001	*
0.69	0.027	0.004	0.001	*
0.7	0.024	0.004	0.001	*
0.71	0.021	0.003	*	*
0.72	0.019	0.002	*	*
0.73	0.017	0.002	*	*
0.74	0.014	0.002	*	*
0.75	0.012	0.001	*	*
0.76	0.011	0.001	*	*
0.77	0.009	0.001	*	*
0.78	0.008	0.001	*	*
0.79	0.007	*	*	*
0.8	0.005	*	*	*
0.81	0.005	*	*	*
0.82	0.004	*	*	*
0.83	0.003	*	*	*
0.84	0.002	*	*	*
0.85	0.002	*	*	*
0.86	0.001	*	*	*
0.87	0.001	*	*	*
0.88	0.001	*	*	*
0.89	0.001	*	*	*
0.9	*	*	*	*

TABLE C.3 *Probabilities of* t *distribution (2-tailed)*

t	df 10	20	30	40	50	100
1	0.341	0.329	0.325	0.323	0.322	0.320
1.1	0.297	0.284	0.280	0.278	0.277	0.274
1.2	0.258	0.244	0.240	0.237	0.236	0.233
1.3	0.223	0.208	0.204	0.201	0.200	0.197
1.4	0.192	0.177	0.172	0.169	0.168	0.165
1.5	0.165	0.149	0.144	0.141	0.140	0.137
1.6	0.141	0.125	0.120	0.117	0.116	0.113
1.7	0.120	0.105	0.099	0.097	0.095	0.092
1.8	0.102	0.087	0.082	0.079	0.078	0.075
1.9	0.087	0.072	0.067	0.065	0.063	0.060
2	0.073	0.059	0.055	0.052	0.051	0.048
2.1	0.062	0.049	0.044	0.042	0.041	0.038
2.2	0.052	0.040	0.036	0.034	0.032	0.030
2.3	0.044	0.032	0.029	0.027	0.026	0.024
2.4	0.037	0.026	0.023	0.021	0.020	0.018
2.5	0.031	0.021	0.018	0.017	0.016	0.014
2.6	0.026	0.017	0.014	0.013	0.012	0.011
2.7	0.022	0.014	0.011	0.010	0.009	0.008
2.8	0.019	0.011	0.009	0.008	0.007	0.006
2.9	0.016	0.009	0.007	0.006	0.006	0.005
3	0.013	0.007	0.005	0.005	0.004	0.003
3.1	0.011	0.006	0.004	0.004	0.003	0.003
3.2	0.009	0.004	0.003	0.003	0.002	0.002
3.3	0.008	0.004	0.002	0.002	0.002	0.001
3.4	0.007	0.003	0.002	0.002	0.001	0.001
3.5	0.006	0.002	0.001	0.001	0.001	0.001
3.6	0.005	0.002	0.001	0.001	0.001	*
3.7	0.004	0.001	0.001	0.001	0.001	*
3.8	0.003	0.001	0.001	*	*	*
3.9	0.003	0.001	0.001	*	*	*

TABLE C.4 Probabilities of the χ^2 distribution

χ^2	df					
	1	2	3	4	5	6
2	0.157	0.368	0.572	0.736	0.849	0.920
2.2	0.138	0.333	0.532	0.699	0.821	0.900
2.4	0.121	0.301	0.494	0.663	0.791	0.879
2.6	0.107	0.273	0.457	0.627	0.761	0.857
2.8	0.094	0.247	0.423	0.592	0.731	0.833
3	0.083	0.223	0.392	0.558	0.700	0.809
3.2	0.074	0.202	0.362	0.525	0.669	0.783
3.4	0.065	0.183	0.334	0.493	0.639	0.757
3.6	0.058	0.165	0.308	0.463	0.608	0.731
3.8	0.051	0.150	0.284	0.434	0.579	0.704
4	0.046	0.135	0.261	0.406	0.549	0.677
4.2	0.040	0.122	0.241	0.380	0.521	0.650
4.4	0.036	0.111	0.221	0.355	0.493	0.623
4.6	0.032	0.100	0.204	0.331	0.467	0.596
4.8	0.028	0.091	0.187	0.308	0.441	0.570
5	0.025	0.082	0.172	0.287	0.416	0.544
5.2	0.023	0.074	0.158	0.267	0.392	0.518
5.4	0.020	0.067	0.145	0.249	0.369	0.494
5.6	0.018	0.061	0.133	0.231	0.347	0.469
5.8	0.016	0.055	0.122	0.215	0.326	0.446
6	0.014	0.050	0.112	0.199	0.306	0.423
6.2	0.013	0.045	0.102	0.185	0.287	0.401
6.4	0.011	0.041	0.094	0.171	0.269	0.380
6.6	0.010	0.037	0.086	0.159	0.252	0.359
6.8	0.009	0.033	0.079	0.147	0.236	0.340
7	0.008	0.030	0.072	0.136	0.221	0.321
7.2	0.007	0.027	0.066	0.126	0.206	0.303
7.4	0.007	0.025	0.060	0.116	0.193	0.285
7.6	0.006	0.022	0.055	0.107	0.180	0.269
7.8	0.005	0.020	0.050	0.099	0.168	0.253
8	0.005	0.018	0.046	0.092	0.156	0.238
8.2	0.004	0.017	0.042	0.085	0.146	0.224
8.4	0.004	0.015	0.038	0.078	0.136	0.210
8.6	0.003	0.014	0.035	0.072	0.126	0.197
8.8	0.003	0.012	0.032	0.066	0.117	0.185
9	0.003	0.011	0.029	0.061	0.109	0.174
9.2	0.002	0.010	0.027	0.056	0.101	0.163
9.4	0.002	0.009	0.024	0.052	0.094	0.152
9.6	0.002	0.008	0.022	0.048	0.087	0.143
9.8	0.002	0.007	0.020	0.044	0.081	0.133
10	0.002	0.007	0.019	0.040	0.075	0.125
10.2	0.001	0.006	0.017	0.037	0.070	0.116
10.4	0.001	0.006	0.015	0.034	0.065	0.109
10.6	0.001	0.005	0.014	0.031	0.060	0.102
10.8	0.001	0.005	0.013	0.029	0.055	0.095
11	0.001	0.004	0.012	0.027	0.051	0.088
11.2	0.001	0.004	0.011	0.024	0.048	0.082
11.4	0.001	0.003	0.010	0.022	0.044	0.077
11.6	0.001	0.003	0.009	0.021	0.041	0.072

continued overleaf

TABLE C.4 (cont.)

χ^2	df 1	2	3	4	5	6
11.8	0.001	0.003	0.008	0.019	0.038	0.067
12	0.001	0.002	0.007	0.017	0.035	0.062
12.2	*	0.002	0.007	0.016	0.032	0.058
12.4	*	0.002	0.006	0.015	0.030	0.054
12.6	*	0.002	0.006	0.013	0.027	0.050
12.8	*	0.002	0.005	0.012	0.025	0.046
13	*	0.002	0.005	0.011	0.023	0.043
13.2	*	0.001	0.004	0.010	0.022	0.040
13.4	*	0.001	0.004	0.009	0.020	0.037
13.6	*	0.001	0.004	0.009	0.018	0.034
13.8	*	0.001	0.003	0.008	0.017	0.032
14	*	0.001	0.003	0.007	0.016	0.030
14.2	*	0.001	0.003	0.007	0.014	0.027
14.4	*	0.001	0.002	0.006	0.013	0.025
14.6	*	0.001	0.002	0.006	0.012	0.024
14.8	*	0.001	0.002	0.005	0.011	0.022
15	*	0.001	0.002	0.005	0.010	0.020
15.2	*	0.001	0.002	0.004	0.010	0.019
15.4	*	*	0.002	0.004	0.009	0.017
15.6	*	*	0.001	0.004	0.008	0.016
15.8	*	*	0.001	0.003	0.007	0.015
16	*	*	0.001	0.003	0.007	0.014

TABLE C.5 F *distribution critical values for* $p = 0.05$

	1	2	3	4	5	6	7	8	9	10
3	10.128	9.552	9.277	9.117	9.013	8.941	8.887	8.845	8.812	8.785
4	7.709	6.944	6.591	6.388	6.256	6.163	6.094	6.041	5.999	5.964
5	6.608	5.786	5.409	5.192	5.050	4.950	4.876	4.818	4.772	4.735
6	5.987	5.143	4.757	4.534	4.387	4.284	4.207	4.147	4.099	4.060
7	5.591	4.737	4.347	4.120	3.972	3.866	3.787	3.726	3.677	3.637
8	5.318	4.459	4.066	3.838	3.688	3.581	3.500	3.438	3.388	3.347
9	5.117	4.256	3.863	3.633	3.482	3.374	3.293	3.230	3.179	3.137
10	4.965	4.103	3.708	3.478	3.326	3.217	3.135	3.072	3.020	2.978
15	4.543	3.682	3.287	3.056	2.901	2.790	2.707	2.641	2.588	2.544
20	4.351	3.493	3.098	2.866	2.711	2.599	2.514	2.447	2.393	2.348
25	4.242	3.385	2.991	2.759	2.603	2.490	2.405	2.337	2.282	2.236
30	4.171	3.316	2.922	2.690	2.534	2.421	2.334	2.266	2.211	2.165
35	4.121	3.267	2.874	2.641	2.485	2.372	2.285	2.217	2.161	2.114
40	4.085	3.232	2.839	2.606	2.449	2.336	2.249	2.180	2.124	2.077
45	4.057	3.204	2.812	2.579	2.422	2.308	2.221	2.152	2.096	2.049
50	4.034	3.183	2.790	2.557	2.400	2.286	2.199	2.130	2.073	2.026
60	4.001	3.150	2.758	2.525	2.368	2.254	2.167	2.097	2.040	1.993
70	3.978	3.128	2.736	2.503	2.346	2.231	2.143	2.074	2.017	1.969
80	3.960	3.111	2.719	2.486	2.329	2.214	2.126	2.056	1.999	1.951
90	3.947	3.098	2.706	2.473	2.316	2.201	2.113	2.043	1.986	1.938
100	3.936	3.087	2.696	2.463	2.305	2.191	2.103	2.032	1.975	1.927

References

Abelson, R.P. (1995). *Statistics as principled argument*. Hillsdale, NJ: Erlbaum.
Aiken, L.S. and West, S.G. (1991). *Multiple regression: testing and interpreting interactions*. London: Sage.
Aitkin, A., et al. (1981). Statistical modelling of data on teaching styles (with discussion). *Journal of the Royal Statistical Society (A)*, *1444*, 148–161.
Aitkin, M., and Longford, N. (1986). Statistical modelling issues in school effectiveness studies. *Journal of the Royal Statistical Society (A)*, *149*, 1–43.
Allison, P. (1999). *Multiple regression: a primer*. Thousand Oaks, CA: Sage.
Atkinson, R.L., Atkinson, R.C., Smith, E.E., Bem, D.J., and Nolan-Hoeksema, S. (1996). *Hilgard's introduction to psychology* (12th ed.). Fort Worth, TX: Harcourt Brace Jovanovich.
Baron, R.M., and Kenny, D.A. (1986). The moderator-mediator variable distinction in social psychological research. *Journal of Personality and Social Psychology*, *51* (6), 1173–1182.
Bennett, N. (1976). *Teaching styles and pupil progress*. London: Open Books.
Berry, W.D. (1993). *Understanding regression assumptions*. Sage University Series Quantitative Applications in the Social Sciences, no. 92. Newbury Park, CA: Sage.
Bollen, K.A. (1989). *Structural equations with latent variables*. New York: John Wiley.
Bollen, K.A., and Long, J.S. (Eds) (1993). *Testing structural equation models*. Newbury Park, CA: Sage.
Browne, M.W. (1984). Asymptotically distribution free methods in the analysis of covariance structures. *British Journal of Mathematical and Statistical Psychology*, *37*, 62–83.
Bryk, A.S., and Raudenbush, S.W. (1992). *Hierarchical linear models*. Newbury Park, CA: Sage.
Chapman, G. et al. (1980). *Monty Python's Big Red Book*. London: Methuen.
Chow, S.L. (1996). *Statistical significance: rationale, validity and utility*. London: Sage.
Clark-Carter, D. (1994). The account taken of statistical power in the British Journal of Psychology. *British Journal of Psychology*, *88* (1), 71–83.
Cochran, W.G. (1965). The planning of observational studies of human populations. *Journal of the Royal Statistical Society*, *128*, 234–255.
Cohen, J. (1962). The statistical power of abnormal-social psychological research: a review. *Journal of Abnormal and Social Psychology*, *65*, 145–153.
Cohen, J. (1988). *Statistical power analysis for the behavioural sciences* (2nd ed.). Hillsdale, NJ: Erlbaum.
Cohen, J. (1994). The Earth is round (p<0.05). *American Psychologist*, *49* (12), 997–1003.
Cohen, J., and Cohen, P. (1983). *Applied multiple regression/correlation analysis for the behavioral sciences* (2nd ed.). Hillsdale, NJ: Erlbaum.
Cohen, P., Cohen, J., Teresi, J., Marchi, M., and Velez, C.N. (1990). Problems in the measurement of latent variables in structural equation causal models. *Applied Psychological Measurement*, *14* (2), 183–195.
Collett, D. (1994). *Modelling survival data in medical research*. London: Chapman and Hall.

Cook, T.D., and Campbell, D.T. (1979). *Quasi-experimentation.* Boston: Houghton Mifflin.
Cudeck, R. (1989). Analysis of correlation-matrices using covariance structure models. *Psychological Bulletin, 105* (2), 317–327.
DeCarlo, L.T. (1997). On the meaning and use of kurtosis. *Psychological Methods, 2* (3), 292–307.
Draper, N.R., and Smith, H. (1981). *Applied regression analysis.* New York: John Wiley.
Dunn, M., and Miles, J.N.V. (1996). A multilevel modelling approach to job search over time. In J. Silvester et al. (Eds.), *British Psychological Society Occupational Psychology Conference Books of Proceedings.* Leicester: British Psychological Society.
Eliason, S.R. (1993). *Maximum likelihood estimation.* Sage University Series Quantitative Applications in the Social Sciences, no. 96. Newbury Park, CA: Sage.
Faul, F., and Erdfelder, E. (1992). *GPOWER: A priori, post-hoc, and compromise power analyses for MS-DOS* [Computer program]. Bonn: Bonn University, Dept. of Psychology.
Fox, J. (1991). *Regression diagnostics.* Sage University Series Quantitative Applications in the Social Sciences, no. 79. Newbury Park, CA: Sage.
Friedman, M., and Rosenman, R.H. (1974). *Type A behaviour.* New York: Knopf.
Gigerenzer, G. (1993). The superego, the ego and the id in statistical reasoning. In G. Keren and C. Lewis (Eds.), *A Handbook for Data Analysis in the Behavioural Sciences: Vol. 1 – Methodological Issues.* Hillsdale, NJ: Erlbaum.
Gigerenzer, G., and Switjink, Z. (1990). *The Empire of Chance: How Probability Changed Science and Everyday Life.* Cambridge: Cambridge University Press.
Godden, D., and Baddeley, A.D. (1975). Context-dependent memory in two natural environments: in land and on water. *British Journal of Psychology, 66,* 325–331.
Goldstein, H. (1995). *Multilevel statistical models* (2nd ed.). London: Arnold.
Gould, S.J. (1981). *The mismeasure of man.* London: Penguin.
Green, S.B. (1991). How many subjects does it take to do a regression analysis? *Multivariate Behavioural Research, 26* (3), 499–510.
Hadi, A.S. (1996). *Matrix algebra as a tool.* Belmont, CA: Duxbury Press.
Hayduk, L.A. (1988). *Structural equation modeling with LISREL: essentials and advances.* Baltimore, MD: Johns Hopkins University Press.
Hayduk, L.A. (1996). *LISREL: issues, debates, and strategies.* Baltimore, MD: Johns Hopkins University Press.
Hoaglin, D.C., and Welsch, R.E. (1978). The hat matrix in regression and ANOVA. *American Statistician, 29,* 73–81.
Hox, J.J. (1995). *Applied multilevel analysis.* Amsterdam: TT-Publikaties.
Hoyle, R.H. (Ed.) (1995). *Structural equation modeling: concepts, issues, and applications.* Newbury Park, CA: Sage.
Hutcheson, G., and Sofreniou, N. (1999). *The multivariate social scientist: introductory statistics using generalised linear models.* London: Sage.
Jaccard, J., Turrisi, R., and Wan, C.K. (1990). *Interaction effects in multiple regression.* Sage University Series Quantitative Applications in the Social Sciences, no. 72. Newbury Park, CA: Sage.
Jaccard, J., and Wan, C.K. (1996). *LISREL approaches to interaction effects in multiple regression.* Sage University Series Quantitative Applications in the Social Sciences, no. 114. Newbury Park, CA: Sage.
Jöreskog, K., and Sörbom, D. (1999). *LISREL 8.30* (Computer program). Chicago: Scientific Software Inc.
Judd, C.M., and McClelland, G.H. (1989). *Data analysis: a model comparison approach.* New York: Harcourt Brace Jovanovich.
Kline, P. (1994). *An easy guide to factor analysis.* London: Routledge.
Kohn, P.M., and MacDonald, J.E. (1991). The survey of recent life experiences: a

decontaminated hassles scale for adults. *Journal of Behavioural Medicine*, *15* (2), 221–236.
Kraemer, H.C., and Thiemann, S. (1987). *How many subjects? Statistical power analysis in research*. Newbury Park, CA: Sage.
Kreft, I., and De Leeuw, J. (1998). *Introducing multilevel modeling*. London: Sage.
Lave, L.B., and Seskin, E.P. (1977). Air pollution and human health. *Science*, *169*, 723–733.
Loehlin, J. (1998). *Latent variable models* (3rd ed.). Hillsdale, NJ: Erlbaum.
Long, J.S. (1983a). *Confirmatory factor analysis: a preface to LISREL*. Sage University Series Quantitative Applications in the Social Sciences, no. 33. Newbury Park, CA: Sage.
Long, J.S. (1983b). *Covariance structure models: an introduction to LISREL*. Sage University Series Quantitative Applications in the Social Sciences, no. 34. Newbury Park, CA: Sage.
Longford, C. (1993). *Random coefficient models*. New York: Oxford University Press.
MacCallum, R.C., and Mar, C.M. (1995). Distinguishing between moderator and quadratic effects in multiple-regression. *Psychological Bulletin*, *118* (3), 405–421.
Marcoulides, G.A., and Schumacker, R.E. (Eds.) (1996). *Advanced structural equation modeling: issues and techniques*. Hillsdale, NJ: Erlbaum.
Maxwell, S.E., and Delaney, H.D. (1990). *Designing experiments and analysis data: a model comparison perspective*. Pacific Grove, CA: Brooks Cole.
Maxwell, S.E., and Delaney, H.D. (1993). Bivariate median splits and spurious statistical significance. *Psychological Bulletin*, *113* (1), 181–190.
McClelland, G.H. (1997). Optimal design in psychological research. *Psychological Methods*, *2* (1), 3–19.
McNemar, Q. (1946). Opinion-attitude methodology. *Psychological Bulletin*, *43* (4), 289–374.
Menard, S. (1995). *Applied logistic regression analysis*. Sage University Series Quantitative Applications in the Social Sciences, no. 34. Newbury Park, CA: Sage.
Mohr, L.B. (1990). *Understanding significance testing*. Thousand Oaks, CA: Sage.
Muthén, B. (1984). A general structural equation model with dichotomous, ordered categorical, and continuous latent variable indicators. *Psychometrika*, *49* (1), 115–132.
Nunnally, J.C., and Bernstein, I.H. (1994). *Psychometric theory* (3rd ed.). New York: McGraw-Hill.
Pagel, M.D., and Lunneborg, C.E. (1985). Empirical evaluation of ridge regression. *Psychological Bulletin*, *97* (2), 342–355.
Pedhazur, E.J. (1982). *Multiple regression in behavioral research: explanation and prediction* (2nd ed.). Fort Worth, TX: Harcourt Brace Jovanovich.
Pedhazur, E.J. (1997). *Multiple regression in behavioral research: explanation and prediction* (3rd ed.). Fort Worth, TX: Harcourt Brace Jovanovich.
Pedhazur, E.J., and Schmelkin, L.P. (1991). *Measurement, design and analysis: an integrated approach*. Hillsdale, NJ: Erlbaum.
Pollard, P. (1993). How 'significant' is 'significance'? in G. Keren, and C. Lewis (Eds.), *A handbook for data analysis in the behavioural sciences: Vol. 1 – Methodological Issues*. Hillsdale, NJ: Erlbaum.
Popper, K.R. (1968). *The logic of scientific discovery*. London: Routledge.
Roberts, M.J., and Russo, R. (1999). *A student's guide to analysis of variance*. London: Routledge.
Rutherford, A. (2000). *Introducing ANOVA and ANCOVA*. London: Sage.
Satorra, A., and Bentler, P.M. (1994). Corrections to test statistics and standard errors in covariance structure analysis. In A. von Eye and C.C. Clogg (Eds.), *Latent variables analysis: applications for developmental research*. Newbury Park, CA: Sage.
Schumacker, R.E., and Lomax, R.G. (1996). *A beginner's guide to structural equation modeling*. Hillsdale, NJ: Erlbaum.

Sedlmeier, P., and Gigerenzer, G. (1989). Do studies of statistical power have an effect on the power of studies? *Psychological Bulletin, 105*.

Shevlin, M. (1995a). Using composite scores in regression/correlation analysis. Paper presented at the Northern Ireland British Psychological Society Annual Conference.

Shevlin, M. (1995b). The effects of unreliablity in regression/correlation analysis. Paper presented at the Psychology Students of Ireland Annual Conference.

Snijders, T.A.B., and Bosker, R. (1999). *Multilevel analysis: an introduction to basic and advanced multilevel modelling.* London: Sage.

SPSS, Inc. (1999). *SPSS* (computer program). Chicago: SPSS, Inc.

Stevens, J.P. (1996). *Applied multivariate statistics for the social sciences* (3rd ed.). Hillsdale, NJ: Erlbaum.

Stevens, S.S. (1946). On the theory of scales of measurement. *Science, 103*, 677–680.

Stevens, S.S. (1951). Mathematics, measurement, and psychophysics. In S.S. Stevens (Ed.), *Handbook of experimental psychology.* New York: John Wiley.

Stigler, S.M. (1980). Stigler's law of eponymy. *Transactions of the New York Academy of Sciences, 39*, 147–157.

Taylor, M.J., and Innocenti, M.S. (1993). Why covariance: a rationale for using analysis of covariance procedures in randomised studies. *Journal of Early Intervention, 17* (4), 455–466.

Thorndike, E.L. (1917). Reading as reasoning: a study of mistakes in paragraph reading. *Journal of Educational Psychology, 8*, 323–332.

Tremblay, T.F., and Gardner, R.C. (1996). On the growth of structural equation modelling in psychological journals. *Structural Equation Modelling, 3* (2), 93–104.

Tukey, J.W. (1977). *Exploratory data analysis.* Reading, MA: Addison Wesley.

Ulrich, R. (1997). FAQ file for STAT-L/sci.stat.consult. http://www.pitt.edu/~wpilib/statfaq/regrfaq.html.

Woodhouse, G. (1996). *Multilevel modelling applications: a guide for users of MLn.* London: Multilevel Models Project, Institute of Education, University of London.

Yerkes, R.M., and Dodson, J.D. (1908). The relation of strength of stimulus to rapidity of habit formation. *Journal of Comparative Neurology and Psychology, 18*, 459–482.

Name index

Abelson, R.P. 116
Aiken, L.S. 168, 187, 191
Aitkin, M. 111, 193
Allison, P. 25
Arbuckle, J. 213
Atkinson, R.C. 116
Atkinson, R.L. 116

Baddley, A.D. 168
Baron, R.M. 188
Bem, D.J. 116
Bennett, N. 193
Bentler, P.M. 200, 212
Bernstein, I.H. 59
Berry, W.D. 112, 135
Bollen, K.A. 115, 135, 200, 206, 215
Bonferroni, C. 50
Bosker, R. 214
Browne, M. 200
Bryk, A.S. 112, 212, 214

Campbell, D.T. 113, 135
Chapman, G. 112
Chow, S.L. 74
Clark-Carter, D. 121
Cochran, W.G. 115
Cohen, J. 25, 26, 39, 57, 74, 120, 121, 133, 134, 135, 191
Cohen, P. 26, 39, 57, 133, 134, 135, 191
Collett, D. 112
Cook, T.D. 113, 135

Darwin, C. 135
De Leeuw, J. 196, 214
DeCarlo, L.T. 65
Delaney, H.D. 41, 50, 191
Dodson, J.D. 137
Draper, N.R. 132
Dunn, M. 111

Eliason, S.R. 157
Erdfelder, E. 121

Faul, F. 121
Fisher, R.A. 115, 116
Fox, J. 112, 131, 135
Friedman, M. 209

Gigerenzer, G. 25, 112, 121
Godden, D. 168
Goldstein, H. 112, 193, 211, 214
Gould, S.J. 116
Green, S.B. 118

Hayduk, L. 214, 215
Hedeker, D. 212
Hoaglin, D.C. 95
Hox, J. 214
Hoyle, R. 200, 214
Hutcheson, G. 45, 164

Innocenti, M.S. 41

Jaccard, J. 168, 187, 191, 215
Jöreskog, K. 207
Judd, C.M. 7, 25

Kenney, D.A. 188
Kline, P. 132
Kohn, P.M. 246
Kraemer, H.C. 120, 135
Kreft, L. 196, 214

Lave, L.B. 134
Loehlin, J. 214
Lomax, R.G. 214
Long, J.S. 215
Longford, N. 111, 214
Lunneborg, C.E. 132

MacCallum, R.C. 187
MacDonald, J.E. 246
Mar, C.M. 187
Marchi, M. 134
Marcoulides, G.A. 215
Maxwell, S.E. 41, 50, 191
McLelland, G.H. 7, 25, 103, 165
McNemar, Q. 200
Menard, S. 159, 164
Mohr, L.B. 18
Muthen, B. 200, 213
Muthen, L. 213

Neale, M. 213
Nolan-Hoeksema, S. 116
Nunnally, J.C. 59

Pagel, M.D. 132
Pedhazur, E. 39, 45, 57, 80, 97, 135, 164
Pollard, P. 25

Rasbash, J. 211
Raudenbush, S.W. 112, 214
Roberts, M. 50
Rosenman, R.H. 209
Russo, R. 50
Rutherford, A. 45, 174

Satorra, A. 200
Schmelkin, L.P. 80

Schumacker, R.E. 214, 215
Sedlmeier, P. 121
Seskin, E.P. 134
Shevlin, M. 134
Smith, E.G. 116
Smith, H. 132
Snijders, T.A.B. 214
Sofreniou, N. 45, 164
Sörbom, D. 207
Spearman, C. 204
Steiger, J. 213
Stevens, J.P. 84
Stevens, S.S. 59
Switjink, Z. 112

Taylor, M.J. 41
Teresi, J. 134
Thiemann, S. 120, 135
Thorndike, E.L. 133
Tukey, J.W. 67, 112
Turrisi, R. 168, 187, 191

Ulrich, R. 38

Velez, C.N. 134

Wan, C.K. 168, 187, 215
Welch, R.E. 95
West, S.G. 168, 187, 191

Yerkes, R.M. 137

Subject index

3D scatterplot 183

adjusted R^2 32
alpha (as type I error rate) 119
ancova 41
AMOS 213
anova
  as simplified regression 40, 45
  as test of R2 33
  for interactions 168
association 113
  direction 114
assumptions 58
  data 61
  levels of measurement 62
attendance
  as predictor 29
attenuation 134
autocorrelation 86, 120

backwards entry 38
balanced design 173
Bonferroni correction 50
box and whisker plot 70
boxplot 70

categorical data, *see* nominal data
categorical variables
  as predictor/independent variable 40
causality 113
  and theory 116
ceiling effect 65
change
  analysis of 52
collinearity 126
  detection 128
  solutions 128
confidence interval
  mean 9
  slope 17

constant (as slope parameter) 10
control
  experimental 31
  statistical 31, 34
Cook's D 96
correlation 20
  calculation 221
  magnitude 25
  multiple 32
  proportion of variance 21
  standardised slope parameter 19
covariance (calculation) 216
cubic transformation 140
curvilinearity 138

deviation coding 50
DfBeta 97
DfFit 98
distance statistics 95
  comparison 96
dummy variable coding 45, 47

effect coding 45
effect size (and power analysis) 120
EQS 212
error
  difference between model and data 2
  measurement 133, 200
  standard deviation 5
exponential (transformation) 83
extraneous variables 31
extroversion 34

floor effect 65
forward entry 38

Gaussian distribution, *see* normal distribution
GPower 121

heteroscedasticity 99
  detection 99
  implications 101
hierarchical data 196
hierarchical models, *see* multilevel models
hierarchical variable entry 34
histograms (to check normality) 67
HLM 212
homoscedascity 99

identification 201
independence assumption 86, 110
indicator coding 46
influence statistics 97
influence statistics (mean) 76
interaction, *see* moderator
intercept (as slope parameter) 10
interval data (level of measurement) 60
inverse transformation 142
isolation 115

joint distributions 84

kurtosis 65, 73
  standard error 74

latent variables 203
least squares models 3
levels of measurement
  assumptions 58
  description 58
leverage 95
line of best fit
linearity
  assumption 105
  non-linearity 137
LISREL 207, 212
log transformation 83, 142
logistic regression 151
  example 151
  purpose 151
logit transformation 154

Mahalanobis distance 96
matrix 223
maximum likelihood 157

mean
  calculation 2
  least squares model 5
  standard error 7
mediation
  complete and partial 187
  described 187
  example 188
method of signatures 117
MIXEDUP 212
MLn 211
MLWin 211
models 1
moderation 165
  <2 categorical variables 173
  2 categorical variables 168
  categorical and continuous variables 174
  two continuous predictors 180
monocausality 28
MPlus 213
multicollinearity, *see* collinearity
multilevel models 193
  algebraic formulation 194
  computer programs 210
  modelling 193
multinomial logistic regression 164
multiple correlation 32
multiple regression 25
multivariate distributions 84
Mx 213

nominal data (described) 59
non-linear analysis 145
non-linear parameters (interpreting) 148
non-linearity 137
normal distribution
  assumption 62, 65
  dealing with departures 67
  detection 67

ordinal data (level of measurement) 59
outliers
  effects on the mean 63
  multivariate 86
  solutions 79

parameter 1
  sample and population 7
path diagram 205

phi correlation 40
point-biserial correlation 40
polynomial logistic regression 164
population (parameter) 7
post-hoc tests 47
power analysis 118
p-p plots 71
principal components analysis 132
probability plots 71

quadratic transformation 138

R^2 32
   test of increase in hierarchical
      regression 36
random slope models, *see* multilevel
      models
ratio data (level of measurement) 60
regression line (calculation) 11
reliability (of measures) 134
residuals
   comparison of types 94
   distribution 85
   mean 3
   regression 15
   studentised 93
   types 90
ridge regression 132
rules of thumb 119

sample sizes 118
scalar 223
SEPath 213
signatures (method of) 117
significance (of slope) 17
simple coding 46
skew 65, 73
   effects 80
   solutions 81
   standard error 74

slope
   equations for simple 10
   as mean difference 43
SPSS 74, 229
standard error
   mean 7
   slope 7
standardisation 18, 78
standardised estimates 18
standardised slope 19
stepwise regression
   description 38
   problems 38, 39
structural equation modelling 198
   programs 212

tolerance 130
transformations (for distributions)
   81
transformations (linearity) 138
true score 204
t-test
   comparing means 42
   parameter estimate 16
type I error
   inflation due to multiple tests 49
   power analysis 119

unbalanced design 173
unreliability (or measures) 134

variable entry (automated) 38
variable entry (hierarchical) 34
variance 21
vector 4, 223
VIF (variance inflation factor) 130

z-scores
   conversion to 18
   detection of outliers 18